A New Dimension of Mathematics with 3D Printing & Design
Grades 6 - 8

A collection of lessons and resources to enhance mathematics instruction through the use of 3D design and printing.

Jill Cochran, Zane Cochran
Mandi Dean, Maggie Sills

Copyright 2017

Table of Contents

Preface 4

Acknowledgements 5

Introduction: About This Book 7

3D Printing Basics 10

Lesson 1: Introduction to 3D Design & Printing 16
SketchUp Tools Checklist Worksheet 25
SketchUp Fish Activity Worksheet 26

Lesson 2: Unit Cube Box Design 27
Design Worksheet 35-36

Lesson 3: Pyramids: Cross Sections and Volume 37
Observation Worksheets 47-48

Lesson 4: Rates and Ratios 49
Observation Tables 59
Proportion Practice Worksheets 60-61

Lesson 5: Cookie Cutter Tessellations 62
Construction Guidelines 73-74

Lesson 6: Cross Sections 75
Shape Review Worksheet 84
Partner Activity Worksheets 85-90

Lesson 7: Tetracubes: 2D vs 3D 91
Sketching Worksheet 99
Clue Sets 100-105

A New Dimension of Mathematics with 3D Printing: **Grades 6 - 8**

Preface

This book contains a variety of lessons developed as we worked in math classrooms to make 3D printing and design more available to a general population of students. As 3D printing has become increasingly popular and accessible, we have seen many schools invest in this technology and utilize it in after school clubs, speciality classes, and as a resource for teachers. However, very few of these efforts have encouraged the use of this technology in mainstream classrooms as a resource in teaching traditional curriculum. We see amazing potential for this application and ways that 3D printing can teach concepts like volume in ways that worksheets or even 3D models cannot.

The opportunities that we've had to facilitate 3D printing in math classrooms have given us experience with a variety of ever-changing software and hardware options. If you are in a position of making these choices, we offer the following advice: the lessons in this curriculum are built with Sketchup, a free design software, which we chose because of the intentional connections that must be made between 2D and 3D objects in the design process. Most activities, however, could also be accomplished with other free programs such as TinkerCAD or Fusion 360.

3D printers have become increasingly reliable and easy to use and there are many good options. As we considered our education-specific challenges we used the following as our criteria for choosing an appropriate printer:
- Visibility - Can students easily see something printing from several sides of the printer or can a webcam easily be attached inside and connected to a projector?
- Durability and size - How easy will it be to move this machine from one classroom (or school) to another?
- Reliability - Will the printer function consistently with an appropriate amount of maintenance?
- Affordability - Is the printer competitively priced against other similar models?
- Materials - Is the 3D printer filament widely available? Does it use a proprietary spool or can any type of filament be used?

Preface

- Support - Is the printer widely used? Are replacement parts readily available?
- Size - Is the build size of the printer reasonable for the types of projects to be printed?
- Connectivity - Can students print to the printer via USB, SD card, or over a network connection?

While this list may seem overwhelming at first, it should be noted that there are a variety of printers that offer a fair balance of many of these criteria. Moreover, the importance of these features may vary based on your particular educational goals, budget, skills, etc.

Acknowledgements

As math education and creative technologies professors interested in community projects, we have had many opportunities to work with teachers, classes and students to develop this curriculum and our ideas about how 3D technologies can enhance our view of mathematics. Throughout these projects we have had several Berry College students that have contributed significantly to the ideas and implementation of this curriculum. This is our team:

Kendra Laney
Mandi Dean
Kat Pugh
Meredith Hopper
Maggie Sills
Julia Gomez
Kaare Nicholas
Chris Whitmire

Additionally, we acknowledge that this curriculum would not have been possible without our partner schools that worked with us to develop lessons and allowed us to work with their students on a variety of projects. The following schools have been invaluable partners:

Berry College Elementary and Middle School
Elm Street Elementary School in Rome City Schools
Coosa Middle School in Floyd County Schools
Rome High School in Rome City Schools

Finally, from funding our first 3D printer to the support of student workers we acknowledge the significant support of Berry College.

Introduction: About This Book

This book is organized with resources that will be helpful for teachers looking to implement 3D printing in classrooms to support mathematical understanding. It is designed with support for even those who are new to the technology while also giving versatile ideas for those that are already comfortable with 3D printing. We start with an introduction to 3D printing for those new to this technology. This is followed by the first student lesson. We highly encourage teachers to teach this lesson as an introduction for students before teaching any of the other lessons in the book. The other lessons can be taught in any order as they complement your regular math curriculum. Where applicable, we have included Common Core math standards from each grade level to facilitate integration with classroom curriculum. Each lesson also includes the following components:

Thinking & Sharing/On Their Own

When it comes to engaging learning experiences, there is a balance between whole class discussion and the sharing of ideas with time to personally interact with the technology or materials. Thus, in each lesson there are sections marked Thinking & Sharing where teachers are encouraged to give instruction and ask thought-provoking questions to facilitate students sharing their own ideas. These sections are sometimes preceded by time for students to explore something On Their Own to motivate the following discussion. At other times the Thinking & Sharing happens before the time to practice On Their Own.

For Their Portfolio

As with other creative endeavors, assessments should be flexible to include a variety of products and ways of demonstrating learning. With each lesson, we have included suggestions for products from the lesson that might make good portfolio pieces. We anticipate that not only will these products help teachers assess student understanding, but that they could also be used to justify this supplementary curriculum to parents and administrators.

In collecting student work for their portfolio, we offer two general suggestions. First, before having students design anything digitally, have them make a plan on paper first. This teaches good design principles, but we have also found that

Introduction: About This Book

students struggle significantly less with computer designs when they have first thought through their ideas using a more familiar medium. Second, we encourage you to establish a uniform and easy way to save students' digital work. This could include be stored online, flash drive, or other method, but you will likely need a method of getting student designs to a 3D printer not connected to their computer.

Going Further

Whether you find that your students love an activity or you would like to have more material on a particular topic, we have included suggestions for additional activities or applications that could naturally follow from the given lesson. Sometimes these suggestions branch into other areas of mathematics and some are follow-up questions directly related to the given activity. In each lesson, these extensions require no additional technical knowledge, but there might be additional mathematical knowledge needed as connections are made to additional topics. We hope that these ideas will inspire you to take our ideas and build on them.

Teacher Talk

From many wonderful conversations with educators, we have developed a rich understanding of the mathematics involved in these lessons. The Teacher Talk section is meant to highlight this rich mathematics for teachers. In this section, we also share common misunderstandings or key concepts involved in each lesson that we have learned from experience. We hope that these insights will help you navigate each lesson well starting the first time that you teach it.

Technology Tips

Each lesson has its own technology challenges. These might include managing and saving student work, displaying a 3D printer for a class to see or any number of other issues that we have had to overcome in a variety of classrooms. The tips are intentionally general for a variety of technology arrangements.

Introduction: About This Book

Online Resources

In addition to everything contained in this book, we have more for you online! Our website www.3DPrintingMath.com, contains printable worksheets, videos referenced in the lessons and other news and updates that you might find helpful as you use these lessons in your classroom. We also welcome feedback on your experience with this curriculum and have provided contact information on the website.

3D Printing Basics

This introduction provides an overview of the basics of 3D printing.

What is a 3D printer?

3D printing is known as an additive manufacturing technique. This means that in order to make an object with a 3D printer, the printer adds material over time, eventually creating a three-dimensional object. This is different from what is known as subtractive manufacturing. A good example of subtractive manufacturing is wood carving. When carving, the carver takes a knife or tool and removes pieces of wood eventually resulting in an object. So 3D printing is an additive process, but how does it actually work?

How does a 3D printer work?

Think about a regular office printer. It prints ink on a piece of paper. A 3D printer does the same thing, except instead of printing one layer of "ink", it does several layers of plastic on top of each other. The 3D printer first prints a single layer of plastic on a platform. Then it prints another layer on top of the first layer. The printer then prints a third layer on top of the second layer, and a fourth layer on top of the third layer, etc. After this repetitive process, you have an object made out of many layers of plastic.

A useful analogy is to think of a 3D printer as a hot glue gun. A hot glue gun takes something cylindrical, a hot glue stick, and melts it so that it is soft and malleable. When this happens, the glue can be extruded out of the glue gun. The 3D printer does the exact same thing to plastic, printing layers of plastic instead of glue.

How do I find or make something to 3D print?

To 3D print something, you either have to scan an object, design your own object using 3D modeling software, or print an existing file. Scanning or designing your own 3D printable object will not be discussed here, but there are a myriad of 3D printable objects that can be found online, that can then be downloaded for free and printed. Thingiverse (www.thingiverse.com) is a one such resource.

3D Printing Basics

Basic Terms

Extruder Temperature: The extruder is the piece of the printer that is heated and pushes out the melted filament. By controlling the temperature of the extruder, someone can optimize the printer's performance for different materials by setting the extruder temperature to the correct temperature for the material being printed.

Filament: The material that is fed into the printer, heated up, and extruded is called filament. Filament can be made of different materials like ABS plastic, PLA plastic, or even flexible materials and often comes on spools in a variety of colors.

Fill and Fill Density: Fill refers to the material that the printer prints inside of the object, and the fill density refers to how much fill is inside the part. Generally, the denser the fill is, the stronger the printed object will be. However, the print will also take much longer and will require more material than if the fill was less dense. In summary, by controlling the fill density, someone can effect the strength of the object, how long the print will take, and how much material is needed to complete the print.

Print Bed Temperature: The print bed temperature refers to the temperature of the platform where the printer prints objects. By controlling this temperature, one is able to affect how much a printed object sticks to the print bed. With larger printed objects, warping can occur if one part of the object cools significantly quicker than another part.

Raft: The raft is a small platform that the 3D printer prints at the beginning of a print. By printing a raft, the printer gives itself a platform to securely print the rest of the object on. Having a raft helps prevent the part from rising up off the platform during a print. Because of this, a platform is much more important to have when using ABS plastic than when using PLA. However, a raft is still helpful when using PLA. One of the biggest drawbacks of using a raft is that someone has to use a knife or paint scraper to remove the raft from the bottom of the printed

3D Printing Basics

object after it has been printed. This can be a tedious process and if not done carefully, can lead to the 3D printed part to break. Note that using a heated bed and Kapton or Teflon tape can be used as an alternative to using a raft.

Speed of print: The speed of the print simply refers to how quickly the printer prints an object. Typically, the slower the print, the higher the print quality will be. The layers will also bond better if the speed is slower. However, having a slow speed does mean the print will take longer. It should also be noted that the longer a print takes, the higher the chance is that something will go wrong and the print will fail. Therefore, lower the speed if you require higher quality, but if quality is not a concern, defer to a quicker speed

STL file: When creating a model file to be 3D printed or looking for a pre-designed 3D printable model, the only files that can be printed are STL files.

Support material: Assume that someone is trying to print a 3D dimensional T. Notice that a T can be created with a vertical rectangle topped with a horizontal rectangle. When the 3D printer begins to print, it will print the vertical rectangle first. Then it will print the horizontal rectangle on top of it. But notice that as the printer starts to print the horizontal rectangle, it will not print it correctly because the parts of the horizontal rectangle that overhang will have nothing to print on. Without support material, a 3D printer would not be able to correctly print any parts that had overhangs. Since overhangs are a common issue, support material can enable an otherwise unprintable object to be printed. Without including supports in the original design, the printer will print support material underneath any overhangs. Once the object is done printing, the support material can easily be peeled away leaving only the finished object. Turning support on/off will control whether or not the printer will use supports to print under overhangs.

Z-resolution: The z-resolution of an object refers to how thick each layer is as the printer creates the object. Smaller z-resolution/layer thickness gives a more detailed printed object.

3D Printing Basics

Materials

There are two materials that you will most likely use while 3D printing. One is ABS plastic and one is PLA plastic. Knowing the differences between these two plastics will help you pick the plastic that meets your project's requirements, and it will also help you pick the right settings for your printer. Please note that both ABS and PLA plastic can absorb moisture from the air which can lead to problems printing with that filament. Because of this, it is better to store the filament in relatively dry place and enclosed container.

ABS

ABS plastic is generally stronger than PLA. It is also more resistant to heat then PLA. This makes it a better pick if the object being printed will be placed under stress or to be used in mechanical parts. Note that ABS requires the printer to have a heated bed.

- Printing temperature: 230-250 degrees C
- Recommended print bed temperature: 80-120 degrees C

PLA

PLA plastic is not as strong as ABS, but it is biodegradable and food safe. This makes it the filament of choice for parts that may come into contact with food. Also, it does not require a heated print bed, unlike ABS.

3D Printing Basics

How to 3D Print an Object

Here are the general steps to follow when 3D printing:

1. Find something to 3D print- Whether designing something yourself, or finding something on Thingiverse, make sure that you save it as an .STL file.
2. In your 3D printer software, load the .STL file you want to print. It works just like opening a document in Microsoft Word.
3. Once your object is loaded, scale it to be the size in which you want it to be. Then rotate it in a way that minimizes overhangs. This will minimize the necessary support material.
4. Place your object on the platform in your 3D printing software. Many programs have a Place or an Automatic Place button that will do this for you. This is necessary because otherwise, the printer will not print your object on the print bed, but instead will try to print it in the air, resulting in a mess.
5. Now that you have your loaded .STL file ready to be printed, make sure that all of the settings within the 3D printer software are correct. This could include selecting your type of material or temperature of the extruder and print bed.
6. Then, much like in Word, click print. This should bring up another screen with print setting options.
7. Customize the print options to fit your needs. Settings here could include things like layer thickness, speed of the 3D print, the density of infill, and the inclusion of infill and/or supports.
8. Click print to start printing your object!
9. As your object begins to print, it is recommended that you keep your eye on the print as the first handful of layers print. The print is most likely to fail while printing these first few layers. After these first layers print successfully, you will only need to check on the print approximately every half hour.
10. Once the print finishes, it is time to remove your print from the printer and do some post-print work on it. The amount of finishing work depends on the material used and whether or not the print has support material or a raft.
11. If there is a raft, remove the print from the print bed with the raft attached. Then by using a paint scraper or knife, peel your object off of the raft.

3D Printing Basics

12. If you have support material, use a knife or your hands to peel off the support. Peeling off support should be relatively easy, but it may be tedious depending on how much support your part needed. Needle-nose pliers are helpful to remove support in hard to reach places.
13. The 3D print is complete at this point, but it can be colored with either markers or paints if desired.

Introduction to 3D Design & Printing

WHAT YOU WILL NEED

- 3D printer
- Sample 3D-printed objects
- .STL file of any object to print
- Method to save student files (e.g. flash drive)

WHAT EACH STUDENT NEEDS

- Computer with SketchUp Make software and .STL Extension
- Tools Checklist handout
- Blank paper
- Pencil

OVERVIEW & PURPOSE

This should be the first lesson with 3D printing regardless of subsequent lessons. In this lesson students use SketchUp software to create and print a three-dimensional object. They will become familiar with the available tools of SketchUp and better understand a 3D printer and how it works.

MATH COMMON CORE STANDARDS

CCSS.MATH.CONTENT.6.G.A.4

Represent three-dimensional figures using nets made up of rectangles and triangles, and use the nets to find the surface area of these figures. Apply these techniques in the context of solving real-world and mathematical problems.

CCSS.MATH.CONTENT.7.G.B.6

Solve real-world and mathematical problems involving area, volume and surface area of two- and three-dimensional objects composed of triangles, quadrilaterals, polygons, cubes, and right prisms.

Introduction to 3D Design & Printing

Have a sample object printing during your lesson and allow students to come and observe the process.

CCSS.MATH.CONTENT.8.G.C.9

Know the formulas for the volumes of cones, cylinders, and spheres and use them to solve real-world and mathematical problems.

THE ACTIVITY
Lesson Duration: 50 minutes

Thinking and Sharing

Get the students excited about 3D printing with an introduction about what a 3D printer is and how it works. If possible, have the 3D printer printing an object during this time and have students either get close to observe it or project a video of it printing. Gauge prior knowledge with questions such as:

- What does it mean for something to be 3D?
- What does a regular printer do (as in a paper printer)?
- Knowing those two things, what do you think a 3D printer does?

Explain that a 3D printer is used to print three-dimensional objects out of plastic. Depending on the printer you have, you can print in just one color or sometimes many colors. Show students the 3D printer and spools of plastic. The plastic is fed into the printer from a spool. There are two types of plastic that are typically used: ABS plastic, which is the same plastic that Legos are made out of, and PLA plastic, which comes from corn. The 3D printer works very much like a hot glue gun. The plastic goes through a very hot nozzle, which melts it and then drops it one tiny bit at a time onto the platform, which is controlled by the computer. The 3D printer prints one layer at a

3D printer filament comes in a variety of colors. The most common types are ABS and PLA.

A New Dimension of Mathematics with 3D Printing: **Grades 6 - 8**

Introduction to 3D Design & Printing

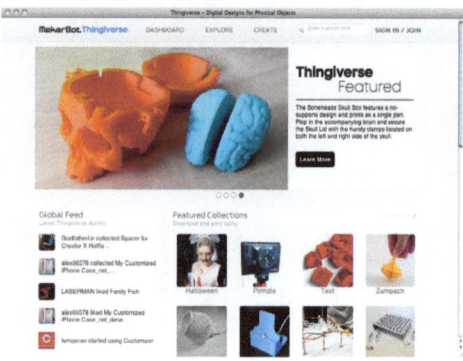

Thingiverse.com is a website where students can browse, download, and share designs that can be 3D printed.

time starting at the bottom. Share and show a variety of objects that can be 3D printed including toys, tools, clothes, artificial limbs, food, etc.

Tell students that anything we want to print must first be created on a computer, and one way to do that is by using a program called SketchUp, which is the software that they will use throughout the lesson. Many 3D designs, however, have already been created by other people, and another way to find things to print is using a website called Thingiverse. Show students Thingiverse.com and have them give a suggestion or two of what to look up. Minecraft characters, Legos, or the school mascot are some possibilities.

Alternatively, this portion of the lesson could be arranged in stations with the following activities: view the 3D printer and spool of plastic and learn about 3D printing, observe objects that have been 3D printed, view a video of something 3D printing, and investigate Thingiverse or similar collection of pictures of things that can be 3D printed.

On Their Own

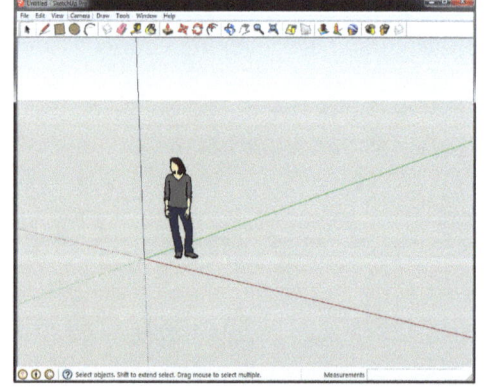

Sketchup Make is a free 3D design software that allows students to design their own objects for the 3D printer.

Students will explore the SketchUp software and the tools that are helpful for creating objects on this software. The students will have 3-5 minutes to explore before the teacher grabs their attention. During this time, verify that all students have the correct program open and an appropriate template selected. We suggest the 3D printing template in millimeters because it sets the scale to an appropriate amount for designing 3D objects that can be printed on most 3D printers.

A New Dimension of Mathematics with 3D Printing: **Grades 6 - 8**

Introduction to 3D Design & Printing

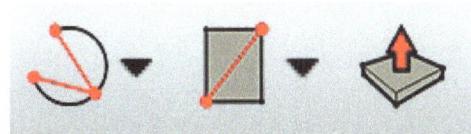

The Shape tool (center) can be used to create rectangles, ellipses, and regular polygons.

Shapes can be set to precise measurements using the Measurements dialog box in the lower right hand size of the screen.

The Push/Pull tool (center) extrudes 2D faces to create 3D solids.

Thinking and Sharing

Once the time is up and you have verified that all students are ready, gather their attention and consider having students put their computers to sleep while you give directions. On a Mac computer, click the Apple symbol in the left top corner and on the drop down there is an option for sleep. On a Windows computer it is part of the Start menu. During this time, project SketchUp and demonstrate how to make a rectangular prism using the tools provided on the software. Go step by step and allow students to ask questions if needed. (For further information on how these tools work, see the introduction.)

Step 1: Introduce the Shape tool. Demonstrate how to "drag" a rectangle (or other shapes) using the cursor.

Step 2: Introduce the Measurement Dialogue Box at the bottom right corner of the screen. This allows the designer to insert precise dimensions for a shape rather than free-handing a design. At this point, the importance of precise measurements is just so that the design pieces "fit" together. Objects can always be re-scaled before 3D printing.

Step 3: Introduce the Push/Pull tool, which allows 2D shapes to become 3D. Like the Shape tool, this can be freehanded or a precise measurement.

Students will explore the other tools on their own as they work through the checklist.

A New Dimension of Mathematics with 3D Printing: **Grades 6 - 8**

Introduction to 3D Design & Printing

On Their Own

From here, have the students return to their own computers and start with a new file.

Pass out the tools checklist where the students will have 10-20 minutes to use all the tools on the checklist and create each of the listed constructions. This is for students to explore the tools at their own pace while ensuring that they have practiced all of the essential tools, such as the Shape tool, Push/Pull tool, and Orbit tool.

Alternatively, have the students choose a particular object to design while they are experimenting with the tools. This can be any object of their choosing, or they may select one from a given list. As they create their object, they will check off each tool as they use it.

As much as time allows, students should learn to input specific dimensions for constructions and use the tools to manipulate objects, growing a more in-depth grasp on the software for future lessons.

Thinking and Sharing

If many students are struggling with the same tool, gather the attention of the class and give step-by-step directions on how to use that tool. If one student feels comfortable with the tool, that student could also demonstrate its use to the class.

On Their Own

By the end of the class, the students will be eager to create their own designs. If time permits, students will

Introduction to 3D Design & Printing

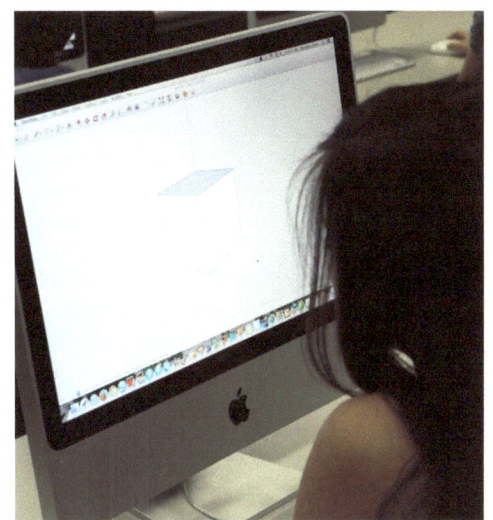

Students can begin creating their own design by constructing a single cube, and then duplicating it.

choose a specific object that they can design themselves. Provide a few options for the students to choose from, such as a birthday cake, robot, or castle. Students will construct their object entirely out of unit cubes, similar to building with blocks in real life. This is the simplest method to provide measurements for the students' first attempt at design. Rather than using unit cubes, more advanced students or classes could design their object using whatever tools they desire.

Go over the steps for creating a unit cube in SketchUp as the students follow along on their own computers.

Step 1: Create a rectangle with a length and width of 1 unit. This is done by typing "1mm,1mm" after beginning to drag out a rectangle.(Note that the SketchUp template is by default in millimeters, but any unit abbreviation can be used to specify the dimensions of the rectangle.)

Step 2: Use the Push/Pull tool to add volume to the square, and enter the same single unit as before so that a cube is created. In other words, begin pulling the square and type "1mm" and enter.

Step 3: Make the cube a "group" so that the shape stays fixed. Then replicate the cube using copy and paste options. To do this, click and drag the cursor over the shape to select the entire cube. The entire cube should be highlighted in blue. Then use these keyboard commands:
- Group - Ctrl + G (or Cmd + G)
- Copy - Ctrl + C (or Cmd + C)
- Paste - Ctrl + V (or Cmd + V)

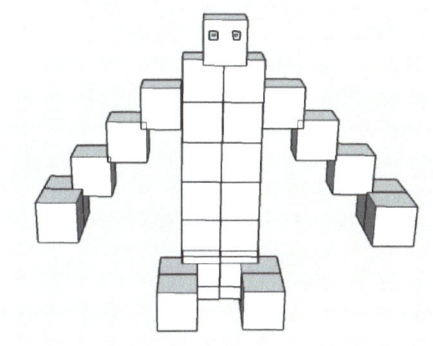

Students can create a wide variety of creative objects using simple cubes in Sketchup.

A New Dimension of Mathematics with 3D Printing: **Grades 6 - 8**

Introduction to 3D Design & Printing

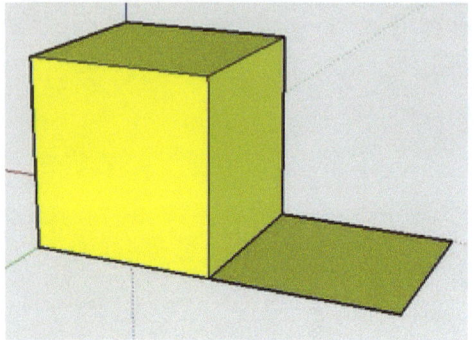

The design above would likely not print because it has a 2D face.

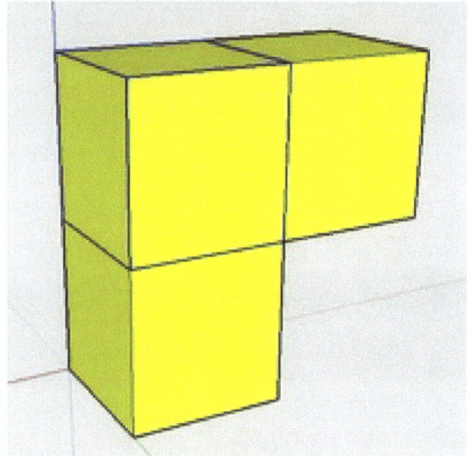

While this design has an overhang, most printers will create support material that will allow it to print without any trouble.

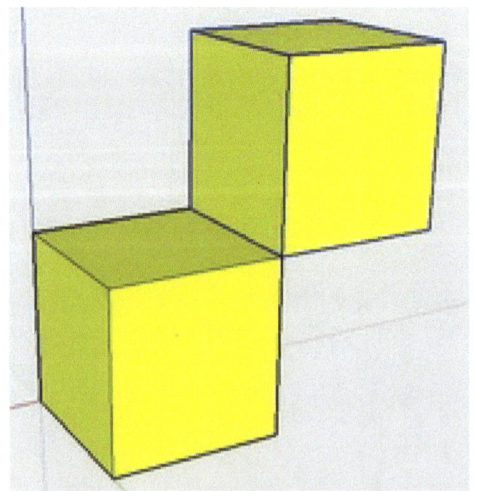

This design would not print well because there is no shared area between the two cubes.

Step 4: Replicate as many cubes as desired. These cubes can then be moved around to create designs. As cubes are moved close to each other they snap together.

Have students ask their partner, neighbors, a "class expert," or the teacher for help if they do not understand how to properly create or replicate a cube.

It can be helpful to give the students five minutes to draw a plan for their design on paper, including specific measurements. Encourage the students to try and think about how their design may look from other angles, trying to get them to see that their designs can have depth as well as height. The students will then be given the rest of class to design their object. These objects may be printed after class for the students to take home. During design time, guide students with questions such as:

- Is your design 3D printable?
- Have you checked your design from multiple viewpoints?

FOR THEIR PORTFOLIO

Before the lesson, the students can complete a pre-assessment. They may be asked to define what a three-dimensional object is or to explain how a two-dimensional object could be turned into a three-dimensional object. At the end of this lesson, the end of the geometry unit, or the end of the 3D printer involvement, the students can again answer these questions to determine how their experiences with 3D printing have influenced their understanding of dimensionality. Also, students' SketchUp designs may be saved.

A New Dimension of Mathematics with 3D Printing: **Grades 6 - 8**

Introduction to 3D Design & Printing

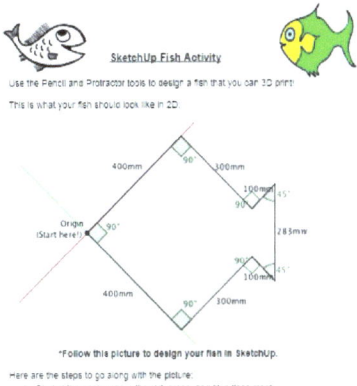

Students can practice their 3D design skills with the Fish Design Activity handout.

When students complete the Fish Design Activity, their design should look like this.

GOING FURTHER

If students proceed quickly through the checklist, they may explore more challenging tools, such as the Rotate tool. They could also have a head start on creating their own design.

Students may also want to explore the designs in 3D Warehouse in addition to Thingiverse. The 3D Warehouse is a site that allows the user to download 3D models directly into their SketchUp file and it can be found under the File menu in SketchUp.

Instead of explicitly telling students how to create a cube, the cube creating activity could be more exploratory, where students discover on their own how to create a proper cube. Lead students to create a 3D object with equal length, width, and height. Ask students to identify this object (a cube) and to distinguish it from a rectangular prism. This can lead into lessons on calculating volume using unit cubes, as well as lessons on properties of geometric solids.

To have students practice using the Protractor and other tools, have them design a fish in SketchUp by following the directions on the "Fish Design Activity" handout. This handout leads students to create a fish using specific lines and angles. This activity could also be used to have students explore different angles, such as acute, obtuse, or right, as well as convex and concave polygons.

TEACHER TALK

- The primary issue in this lesson that could cause

A New Dimension of Mathematics with 3D Printing: **Grades 6 - 8**

Introduction to 3D Design & Printing

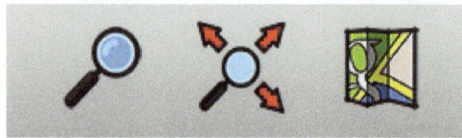

The Zoom Extents tool (center) automatically zooms and centers the design in the window.

The Orbit tool allows students to inspect their designs from a variety of angles to check it for errors before printing.

differing levels of comprehension is students' lack of focus. Often, students are excited to explore this new technology on their own, but this excitement may lead to unsupervised computer activity and disruptive behaviors pulling attention away from the lesson being taught.

- Emphasize the distinction between 2D and 3D designs. A 3D object will print because it has volume; a 2D object will not print because it does not have volume, so there is nothing for the 3D printer to print.
- The layers of a 3D print can lead into lessons on cross sections. For instance, students might not be able to understand why a cylinder contains a circle, but by watching the object being created layer by layer the student is physically able to see each layer. If students are struggling to enter specific measurements, this portion of the lesson could be skipped or saved for another day. However, the ability to make precise designs by entering specific measurements will likely be useful to the students in future 3D printing lessons.

TECHNOLOGY TIPS

- Remember that if a student's design seems to disappear, it may just be incredibly small. Use the Zoom Extents tool to zoom in perfectly on current creations in SketchUp.
- Students should be encouraged to view their design from multiple angles to ensure that it is actually 3D and that all components of the design are actually connected to each other so that it will print properly. The 2D screen representation of a 3D design can often hide these issues.

A New Dimension of Mathematics with 3D Printing: **Grades 6 - 8**

Introduction to 3D Design & Printing

SKETCHUP TOOLS CHECKLIST - WORKSHEET   Name _____

Check off each construction once you make it:
___ Circle
___ Rectangle
___ Triangle

___ Circle with radius 30mm
___ Rectangle that is 20 x 40mm

___ Cylinder
___ Rectangular prism
___ Cube

Make sure you have used each of these tools. Check off each tool once you use it.

Eraser   Paint   Pan   Move   Orbit   Zoom   Extents

A New Dimension of Mathematics with 3D Printing: **Grades 6 - 8**

Introduction to 3D Design & Printing

SKETCHUP FISH ACTIVITY - WORKSHEET   Name _____

Use the Pencil and Protractor tools to design a fish that you can 3D print!

This is what your fish should look like in 2D:

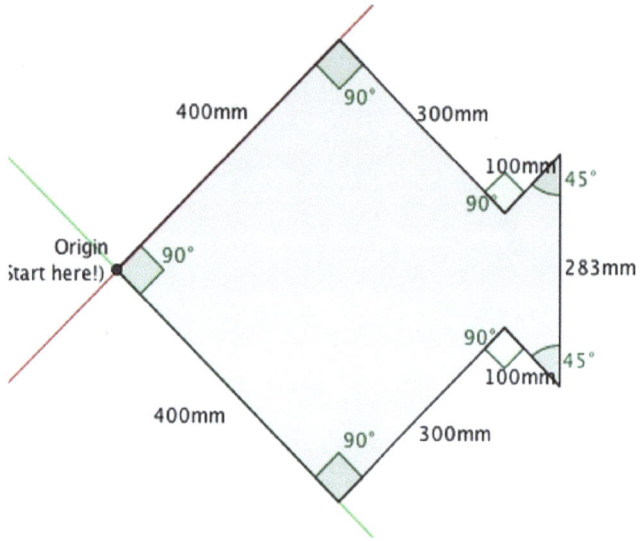

*Follow this picture to design your fish in SketchUp.

Here are the steps to go along with the picture:
1. Start at the origin (where the red, green, and blue lines meet).
2. Make a line 400mm long on the green axis.
3. Measure a 90° angle from the line you just made (using the Protractor)
4. Make a line 300mm long on the guideline you just made.
5. Measure a 90° angle from the line you just made.
6. Make a line 100mm long on the guideline you just made.
7. Measure a 45° angle from the line you just made.
8. Make a line 283mm long on the guideline you just made.
9. Measure a 45° angle from the line you just made.
10. Make a line 100mm long on the guideline you just made.
11. Measure a 90° angle from the line you just made.
12. Make a line 300mm long on the guideline you just made.
13. Connect the line you just made back to the origin.

A New Dimension of Mathematics with 3D Printing: **Grades 6 - 8**

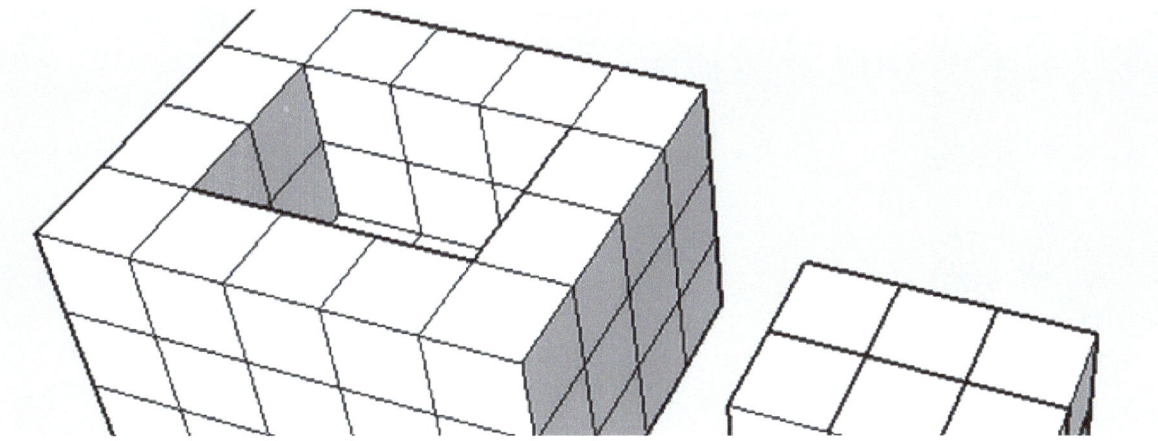

Unit Cube Box Design

WHAT YOU WILL NEED

- 3D printer
- Sample 3D-printed boxes (one with the capacity to hold 12 unit cubes, and one with an outer volume of 12 cubic units)
- Method to save student files (e.g. flash drive)

WHAT EACH STUDENT NEEDS

- Computer with SketchUp Make software and .STL Extension
- 12 unit cubes (1cm^3)
- Guiding worksheet
- Pencil

OVERVIEW & PURPOSE

In this lesson students will design a box to hold twelve cubes. This activity emphasizes the concept of volume by packing a prism with unit cubes. It also introduces students to the real-world design process, where designs are based on measurements of real-world objects.

MATH COMMON CORE STANDARDS

CCSS.MATH.CONTENT.6.G.A.2

Find the volume of a right rectangular prism with fractional edge lengths by packing it with unit cubes of the appropriate unit fraction edge lengths, and show that the volume is the same as would be found by multiplying the edge lengths of the prism. Apply the formulas V = l w h and V = b h to find volumes of right rectangular prisms with fractional edge lengths in the context of solving real-world and mathematical problems.

CCSS.MATH.CONTENT.7.G.B.6

Solve real-world and mathematical problems involving area, volume and surface area of two- and three-dimensional objects composed of triangles, quadrilaterals, polygons, cubes, and right prisms.

A New Dimension of Mathematics with 3D Printing: **Grades 6 - 8**

Unit Cube Box Design

In this activity, students will design a container to hold 12 standard unit cubes.

THE ACTIVITY
Lesson Duration: 50 minutes

Thinking and Sharing

In this lesson, students will be given the opportunity to design an object based off of dimensions of another object. To introduce this concept, share with students that real-world designers often base their creations on measurements of real-world objects.

Ask students to give their definition of volume. As they share ideas, help clarify and summarize so that the following idea is expressed: volume is the amount of 3D space an object occupies. If desired, remind students that objects must be 3D in order to 3D print; that is, an object must have volume in order for the 3D printer to recognize the design and print it out.

Show students a single unit cube with dimensions 1cm x 1cm x 1cm. Alternatively, any sized unit cube will work. Emphasize that the volume of this single cube is one cubic centimeter. Show some pictures and/or physical designs that are made solely out of unit cubes, asking the students what the volume of these designs are. The students should recognize they can find the volume of a design by adding up the total number of unit cubes used to create the design.

An example of a student's container design for 12 unit cubes.

Have students open SketchUp on their computers. Hand out twelve physical unit cubes to each student, or to each group of students if there are not enough cubes for all students to have their own. Assign the students this task:

A New Dimension of Mathematics with 3D Printing: Grades 6 - 8

Unit Cube Box Design

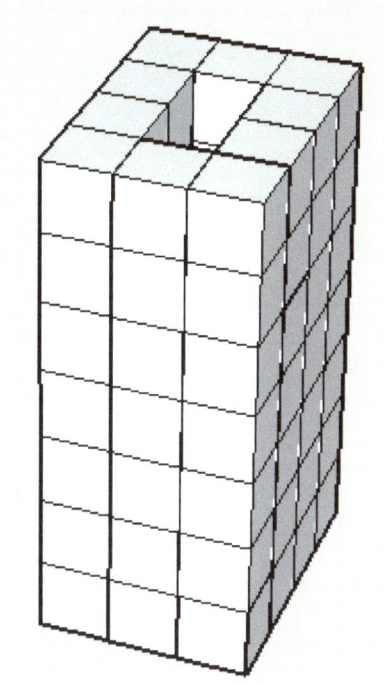

An alternative solution for designing a container to hold 12 unit cubes.

Design a box in SketchUp that will hold all twelve of your cubes perfectly. You can only use unit cubes in SketchUp to build your box, so that your design is made out of lots of unit cubes. The box does not need to have a lid. This box will be 3D printed for you to keep. Once it is printed, you can see if your twelve cubes really fit into your box!

Alternatively, the students could imagine that their unit cubes are dice and that they are constructing a box to hold their dice.

Pass around a previously printed box for students to examine. Additionally, project pictures of some boxes created out of cubes so that students can get an idea of what they should design. It may be difficult for students to understand that the box is supposed to hold unit cubes and also be made out of unit cubes. Make sure to clarify this with the students.

Explicitly go over the steps for creating a unit cube in SketchUp as the students follow along on their own computers.

Step 1: Create a rectangle with a length and width of 1 unit. For a centimeter cube, the rectangle dimensions should be entered as "1cm, 1cm."

Step 2: Use the Push/Pull tool to extrude the square, and enter the same measurement as before (for example, 1cm) so that a cube is created.

Step 3: Make the cube a "group" so that the shape stays fixed. Then replicate the cube using copy and paste

A New Dimension of Mathematics with 3D Printing: **Grades 6 - 8**

Unit Cube Box Design

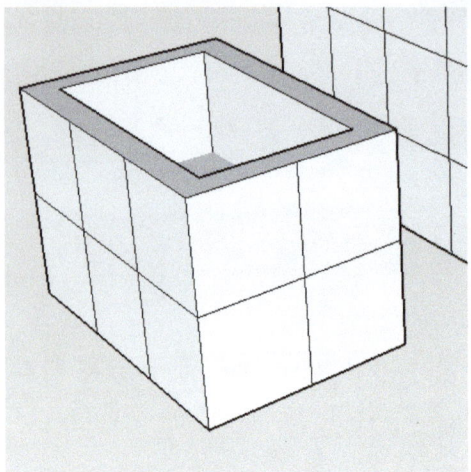

This design did not follow the desired method for creating the container. Note the lack of unit cubes in the design.

options. To do this, drag the cursor over the shape and let it go to select the entire cube. The entire cube should be highlighted in blue. Then use these keyboard commands:

- Group - Ctrl + G (or Cmd + G)
- Copy - Ctrl + C (or Cmd + C)
- Paste - Ctrl + V (or Cmd + V)

Step 4: Replicate as many cubes as desired. These cubes can then be moved around to create designs. Have students ask their partner, neighbors, a "class expert," or the teacher for help if they do not understand how to properly create or replicate a cube.

On Their Own

Once students have created a proper unit cube and are able to replicate it, they may begin designing their box. Provide students with a guiding worksheet with space for them to draw an initial plan of their box before they begin designing. This plan should include specific measurements. Students often want to skip this step, but emphasize that planning will save time and frustration later. This is an important part of the design process.

As students design, ask guiding questions such as:

- How do you know your box will hold twelve unit cubes?
- How much volume will your box hold? Is this the same as the total volume of the box?
- Does your box need to be bigger or smaller to hold all your cubes perfectly?

A New Dimension of Mathematics with 3D Printing: Grades 6 - 8

Unit Cube Box Design

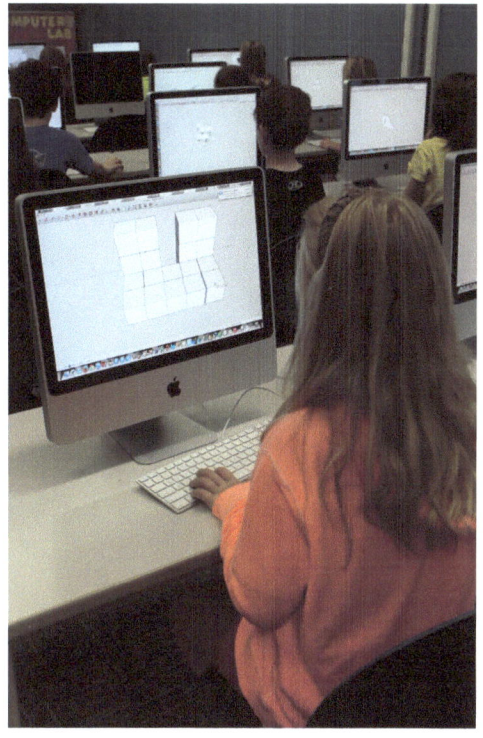

If students are struggling to design a box of correct measurements, have them create a physical rectangular prism out of their twelve unit cubes, and then instruct them to build a box that can hold that prism.

Alternatively, if the students are likely to struggle with the directions for this activity, start with a demonstration with four unit cubes. Pass around different 3D printed boxes that will hold four cubes, as well as a few boxes that are too small or too large to perfectly hold four unit cubes. Then the students can extend the idea to design a box for twelve unit cubes.

If time allows, encourage students to verify that their design will in fact hold twelve unit cubes. This can be done by using the Tape Measure tool in SketchUp to measure the inside of their box. Students could also create twelve additional cubes in SketchUp and fit those cubes into their box to verify the dimensions. For example, in the picture below, the rectangular prism of twelve cubes should fit perfectly into the box. The box contains a volume of twelve cubic units.

Students may find through this process that their dimensions are incorrect and that they need to try again or update their design.

Thinking and Sharing

Students will finish designing at different times. For those who finish quickly, allow them to personalize their design. Alternatively, these students could create another box with different dimensions than their first box, but still containing twelve cubes.

A New Dimension of Mathematics with 3D Printing: Grades 6 - 8

Unit Cube Box Design

As more students begin to finish, have students compare their boxes to someone else's box. If possible, match students together who designed boxes with different dimensions. The students should check the validity of the other student's design, taking into account that both of their boxes will contain a volume of twelve cubic units despite the differing dimensions. Another instructional method may be to call on students individually to present their designs to the class.

Students' designs cannot all be printed during this class for them to verify right away if their designs will hold their cubes, so it is helpful to have a few demo boxes already printed that the students can examine. A few of the demo boxes should have correct dimensions, and many students may recognize that their designs match one of those boxes. Additionally, print a few demo boxes that have incorrect dimensions. In particular, print at least one box with an outer volume of twelve cubic units, which may be a common design mistake among students. By examining this box, students will easily see that such a box is too small to hold all twelve of their cubes.

On Their Own

3D print the students' boxes and hand them back. The students can then physically determine whether their twelve cubes will fit into their box. Have students reflect on this task by asking these questions on a worksheet:

- Did your design turn out the way you expected? Why or why not?
- Does your box hold twelve unit cubes?

A New Dimension of Mathematics with 3D Printing: **Grades 6 - 8**

Unit Cube Box Design

Advanced students can practice designing containers for unit cubes arranged in creative ways.

- What is the volume of the inside of your box? How do you know?
- Are you surprised that your box and someone else's box both hold twelve unit cubes, even though your boxes look very different? Why were you surprised or not surprised?
- How can two boxes with different sizes hold the same amount of volume?

Thinking and Sharing

After students have recorded their observations and conclusions, have them share some of their findings.

FOR THEIR PORTFOLIO

The students' box designs should perfectly contain twelve unit cubes. The students should be able to identify this volume as twelve cubic units. Students should record their discoveries on the worksheet.

GOING FURTHER

- Instruct students to create another box that holds twelve unit cubes with different dimensions than the first box they designed.
- Advanced students could create a box without using solely unit cubes to build. They could construct a rectangle as the base of their box, pull the box up to the desired height, and then erase an inner rectangle of the box to contain the twelve unit cubes. This design requires the students to plan their dimensions more in advance and to possess competency in entering specific measurements into the measurement dialogue box.
-

Unit Cube Box Design

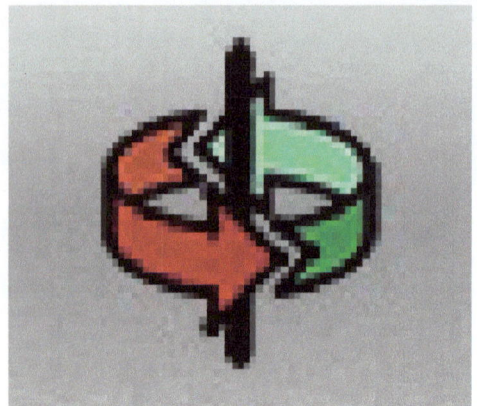

The orbit tool can help students arrange blocks closely together to avoid printing errors.

- If a student wanted to extend this activity, they could design a lid for their box. They could also design a box that is not a rectangular prism.

TEACHER TALK

- Twelve is a useful number of unit cubes, because there are many different potential dimensions of a box containing a volume of twelve cubic units.
- It may be difficult for students to understand that the box is supposed to hold unit cubes and also be made out of unit cubes. Make sure to clarify this with the students.
- At first, many students may create a box with an outer volume of twelve cubic units, which will be too small since the box's walls will take up some of that space. By physically examining a box with an outer volume of twelve cubic units, students will easily see that such a box is too small to hold all twelve of their cubes. Make sure to distinguish between the volume of the box, as calculated by the outer dimensions of the box, and the volume that the box can contain. Because the walls of the box must have thickness, the outer volume will be greater than the volume that the box can contain. This is a common real-world design consideration.

TECHNOLOGY TIPS

As students are building with the cubes, they may notice that the cubes do not always go exactly where they want them to, due to the 2D screen representation of 3D objects. It often helps to move the screen around or orbit to another angle until the cube can be moved to the desired location.

A New Dimension of Mathematics with 3D Printing: **Grades 6 - 8**

Unit Cube Box

DESIGNING WORKSHEET

Name _____

Draw a plan for your box in the space below. Include specific measurements for the dimensions of your box.

Design your box in SketchUp. Then answer these questions:

1. How do you know your box will hold twelve unit cubes?

2. How much volume will your box hold? Is this the same as the total volume of the box? Why or why not?

3. Does your box need to be bigger or smaller to hold all your cubes perfectly?

A New Dimension of Mathematics with 3D Printing: **Grades 6 - 8**

Unit Cube Box

DESIGNING WORKSHEET   Name _____

After you have received your 3D printed box, answer these questions:

1. Did your design turn out the way you expected? Why or why not?

2. Does your box hold twelve unit cubes? Could your design be improved?

3. What is the volume of the inside of your box? How do you know?

4. Are you surprised that your box and another box both hold twelve unit cubes, even though your boxes look very different? Explain your reasoning.

5. How can two boxes with different sizes hold the same amount of volume?

A New Dimension of Mathematics with 3D Printing: **Grades 6 - 8**

Pyramids: Cross Sections & Volume

WHAT YOU WILL NEED

- 3D printer
- Sample 3D-printed objects
- .STL file of pyramid and rectangular prism to print
- Pre-printed hollow rectangular prism and pyramid with same base and height
- Sand, rice, or water
- Method to save student files (e.g. flash drive)

WHAT EACH STUDENT NEEDS

- Computer with SketchUp Make software and .STL Extension
- Pyramid Activity handout
- Pencil

OVERVIEW & PURPOSE

This lesson has students compare the cross sections of a pyramid with the cross section of a rectangular prism with the same base and height. Students will also develop the formula for the volume of a pyramid by comparing pyramids with rectangular prisms, and they will apply this formula to pyramids that they create in SketchUp that can be 3D printed.

MATH COMMON CORE STANDARDS

CCSS.MATH.CONTENT.6.G.A.3

Describe the two-dimensional figures that result from slicing three-dimensional figures, as in plane sections of right rectangular prisms and right rectangular pyramids.

CCSS.MATH.CONTENT.7.G.B.6

Solve real-world and mathematical problems involving area, volume and surface area of two- and three-dimensional objects.

CCSS.MATH.CONTENT.8.G.C.9

Know the formulas for the volumes of cones, cylinders, and spheres and use them to solve real-world and mathematical problems.

A New Dimension of Mathematics with 3D Printing: **Grades 6 - 8**

Pyramids: Cross Sections & Volume

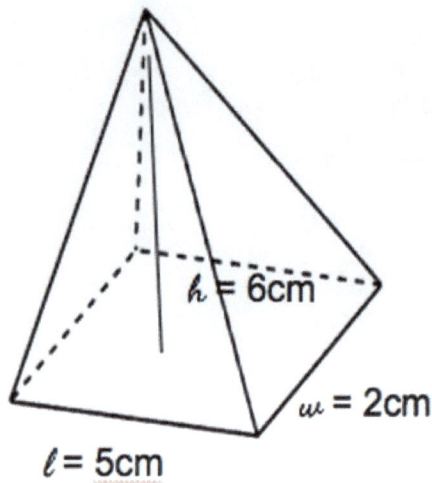

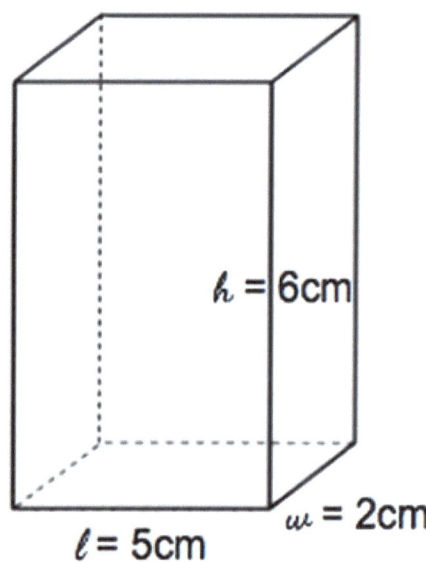

Ask students to compare these two solids to prompt them in thinking about the properties of pyramids.

THE ACTIVITY
Lesson Duration: 50 minutes

Thinking and Sharing
Download the .STL file of a pyramid and a rectangular prism into the 3D printer software. At the beginning of class, start the print.

This lesson will introduce students to the concept of exploring pyramids. To develop student interest for this topic, show some pictures of pyramids in the real world as well as a picture comparing a pyramid and rectangular prism with the same base and height.

Ask the students how these two geometric solids compare and contrast. Some answers the students might come up with are:

- The bases are the same shape on both solids.
- The height is the same.
- The number of faces are different.
- The shapes of the faces are different. The rectangular prism has rectangular faces, while the pyramid has some triangular faces as well as a rectangular base.
- The prism has two bases and a pyramid has one base.
- The side faces of the pyramid meet up at a point called a vertex.

This lesson will have students compare pyramids and rectangular prisms by examining the cross sections of these solids. Review or define the term cross section, which is a 2D shape that results from cutting a 3D solid

A New Dimension of Mathematics with 3D Printing: **Grades 6 - 8**

Pyramids: Cross Sections & Volume

3D Print some sample pyramids before class and allow students to examine them.

with a plane. With both the rectangular prism and the pyramid, ask students: If we cut horizontal to the base, what shape are the cross sections of this solid?

On Their Own

Give each student a "Pyramid Activity" handout. Have students gather around the 3D printer a few at a time to observe the print of the pyramid and rectangular prism taking place. Alternatively, a webcam could be installed in the printer so that live video of the print could be projected in front of the class for all the students to watch at once.

As the students observe the solids being printed, they should record their observations in the tables provided on the front page of the handout. In these tables, the students will draw the first layer, the middle layer, and the top layer that the 3D printer has printed or will print for both of the solids. These layers represent various cross sections of the two solids that are horizontal to the base.

Thinking and Sharing

After students have drawn their observations, they should compare their pictures with a partner, explaining how they came up with their pictures. Then bring the entire class back together to share their observations. Start with questions such as:

- What will the very top layer of the pyramid look like?
- Were the bottom layers for the two solids the same or different? What about the middle layer? Top layer?

The students should have observed that the first layers

A New Dimension of Mathematics with 3D Printing: **Grades 6 - 8**

Pyramids: Cross Sections & Volume

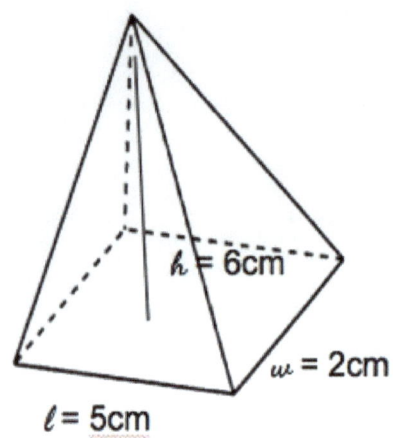

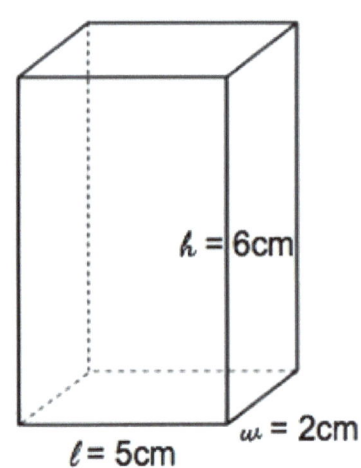

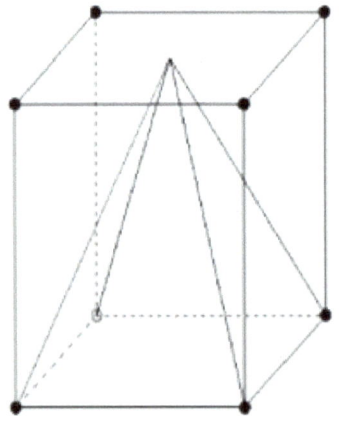

for both of the solids were the same, since their bases have the same dimensions. However, as more and more rectangular layers were added to both solids, the rectangular cross sections of the pyramid shrank in size, while the rectangular cross sections of the prism remained the same. Eventually, the top layer of the pyramid would be a rectangle so small that it would only be a single point, forming the top vertex of the pyramid. Thus the students can distinguish between these two solids by comparing their cross sections.

Next, extend these ideas to volume. Ask students which of the two solids they think would have the greater volume. Students should conclude that since the cross sections of the pyramid are either the same or smaller than the cross sections of the prism, the pyramid would have a smaller volume than the prism. This can be phrased in terms of the amount of plastic necessary to print each solid. Each layer of the pyramid requires a less than or equal amount of plastic as the same layer of the prism. There are the same number of layers for each solid. The students may also reason that a pyramid would fit inside a rectangular prism with the same base and height, so the volume of the pyramid must be smaller than the volume of the prism.

Using the picture above, either have the students calculate the volume of the rectangular prism, or inform them that the volume is 60cm^3. Knowing this information, ask the students to estimate the volume of the pyramid, and to explain their reasoning. To investigate whether the students' hypotheses are correct, perform an experiment with a hollow, pre-printed pyramid and rectangular prism, both with the same dimensions for the base and height.

A New Dimension of Mathematics with 3D Printing: **Grades 6 - 8**

Pyramids: Cross Sections & Volume

It would be beneficial to have pictures of a hollow, printed prism and pyramid, but not necessary.

Have one or two students come to the front of the room to perform the experiment. Have them fill the pyramid with sand, rice, or water, and then pour out these contents to the rectangular prism. They should continue this process until the rectangular prism is completely full.

Alternatively, print enough pyramids and prisms for each student, or group of students, to have their own. Then each student or group of students could perform this activity perhaps with different pairs of prisms and pyramids.

The students will observe that it takes three times the volume of the pyramid to fill the rectangular prism. The students should know that the volume of a rectangular prism is given by $V=B*h$, where B is the area of the base and h is the height of the solid. Using this information, give students a minute to think on their own what the formula for the volume of a pyramid would be, assuming that the pyramid has the same base and height as a rectangular prism. Then students will share their formulas as a class. From the experiment, students should realize that the formula for the volume of a pyramid is given by $V=(B*h)/3$.

On Their Own

Once the students have developed this formula they will apply this formula by creating their own pyramid and rectangular prism in SketchUp. These solids can be any dimensions, but they must have the same base and height as each other. Start by showing students a picture of a pyramid from the top view, so that it looks like the

Pyramids: Cross Sections & Volume

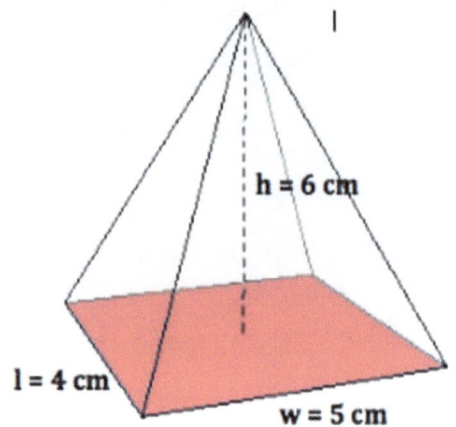

picture below:

Have the students brainstorm ways to make a pyramid in SketchUp. Students may conjecture that they should create a rectangular prism in SketchUp and then remove part of it to turn it into a pyramid. This is a strong idea, but it is difficult to carry out in the software. The students will likely not be able to come up with the easiest way to create a pyramid in SketchUp, so they should be led through the steps. Even though students will not usually arrive at this procedure on their own, it is worth having them explore possible methods as the comparison between the solids leads to a deeper understanding of three dimensional solids and volumetric relationships.

To create a pyramid, first create a rectangle using the Shape tool as a base for the pyramid and typing in the desired dimensions. Next, use the Line tool the create diagonal lines between both opposite vertices of the rectangle. Point out that this creation looks identical to the top view of the pyramid from before. All that is left to do is to make the shape three-dimensional. This is done using the Move tool, which allows the user to pull the shape up from the middle point where the diagonals intersect. Click on this middle point and then pull it up using the Move tool and type in desired height. This point becomes the vertex at the top of the pyramid.

After this instruction, have students open and save a new SketchUp file. Each student will design both a pyramid and prism with the same base and height for comparison. The pyramid they create in this new file may be 3D printed and given back during a later class for them to keep. The

A New Dimension of Mathematics with 3D Printing: **Grades 6 - 8**

Pyramids: Cross Sections & Volume

students will record the dimensions of their two solids on the back side of the "Pyramid Activity" handout. This helps reinforce that the solids have the same base and height. Some students may design their solids and then use the Ruler tool to determine the dimensions. Other students may write down their dimensions before designing in SketchUp, and then use the Measurement input box to enter specific measurements for their solids. Both of these methods are valid approaches. Once the students have finished designing, they will calculate the volume of both their pyramid and prism using the created formulas and record these values on the handout. They will then compare the two values. The volume of the prism should be three times larger than the volume of the pyramid.

If there is extra time in class, or if a student finishes their design early, they may embellish their pyramid with creative designs. They could also download a pyramid from the 3D Warehouse in SketchUp to explore.

Thinking and Sharing

To conclude the lesson, remind students that pyramids and rectangular prisms are distinguishable by their cross sections. As they printed, the two solids started out with the same base, but the cross sections of the pyramid got smaller and smaller while the cross-sections of the prism stayed the same. This concept also helps in the comparison of the volume for a pyramid to the volume of a rectangular prism with the same base and height, which is the same except it is divided by three. Point to the formulas for both the volume of a rectangular prism and volume of a pyramid, which should clearly reflect all

A New Dimension of Mathematics with 3D Printing: **Grades 6 - 8**

Pyramids: Cross Sections & Volume

the discoveries from the lesson.

For an assessment, have the students complete a quick exit slip before the end of class. This could be simply having them calculate the volume of a pyramid, such as the one below.

FOR THEIR PORTFOLIO

- Students' drawings of cross sections on the handout
- Students' calculations of volume on the handout
- Students' pyramid designs in SketchUp
- Exit slip

GOING FURTHER

- A similar lesson can be conducted with cones and cylinders rather than pyramids and rectangular prisms. The volume of a cylinder is three times the volume of a cone with the same base and height.
- The students' observations of the cross section of the pyramid printing can be used to explore oblique pyramids. Simultaneously 3D print a right pyramid and an oblique pyramid, both with the same base and height measurements. Alternatively, the students could just observe a picture of these two solids. Ask students which one they think has a greater volume. As they ponder this question, have them focus on each cross section, or layer that the 3D printer is printing. Students should observe that as the 3D printer prints both of these pyramids, each layer will be exactly the same size; the oblique pyramid's layers are skewed as compared to the right pyramid, but they still have the exact same area. This observation about the layers of the 3D print should help students to realize that the

Pyramids: Cross Sections & Volume

volume for the two pyramids is the same. Thus, the volume for a pyramid can always be calculated by using the height, regardless of whether the height is directly in the center of the pyramid.
- This lesson can be connected to dilations, similarity, and proportions. Provide students with a picture of a famous real-life pyramid, and include the real-world measurements of the pyramid. The students must scale down this pyramid until it is a size that is 3D printable. By explaining why their scaled down pyramid is similar to the real pyramid, students will be defining similarity in the context of a real-world scenario.

TEACHER TALK

- This lesson helps to relate the concept of volume with the concept of cross sections.
- This lesson is strong because it provides students with multiple visuals for calculating the volume of a pyramid before they even get to the formula. The formula seems to emerge as a natural progression from their observations. First, the students are able to observe the cross sections of a pyramid as they watch it 3D print. Then they are able to figure out how the volume of a pyramid compares to a rectangular prism using physical models. Finally, the students design these solids for themselves to calculate the volumes.

TECHNOLOGY TIPS

- Creating pyramids can sometimes be tricky for students. The students will likely be used to using the Push/Pull tool to make their 2D objects 3D. However,

Pyramids: Cross Sections & Volume

this only works to pull up a face or flat surface. When creating a pyramid, we want to pull up only the vertex, so we must use the Move tool rather than the Push/Pull tool. This allows us to pull the shape up from a single point.

- Even when they are using the Move tool, pulling the pyramid up into three dimensions can present problems for students, and it may take them a few tries to get it right. SketchUp sometimes moves the vertex horizontally, along the same plane as the rectangular base, rather than vertically, even though the vertex may appear to be moving vertically. Students should orbit around their pyramid to check that their design is correct. Also, encourage students to be persistent and try to pull the vertex up from different viewing angles until it works.

Pyramids

OBSERVATION TABLES

Name _____

Draw cross-sections of the two solids being printed.

Pyramid		
First layer:	Middle layer:	Top layer:

Rectangular Prism		
First layer:	Middle layer:	Top layer:

A New Dimension of Mathematics with 3D Printing: **Grades 6 - 8**

Pyramids

OBSERVATION WORKSHEET

Name _____

Design Challenge: Design a rectangular prism and a pyramid with the same base and height.

Record the measurements of your recrtangular prism:

Length=_____   Height= _____   Width= _____

Volume of prism=_____

Record the measurements of your pyramid:

Length=_____   Height= _____   Width= _____

Volume of pyramid=_____

A New Dimension of Mathematics with 3D Printing: **Grades 6 - 8**

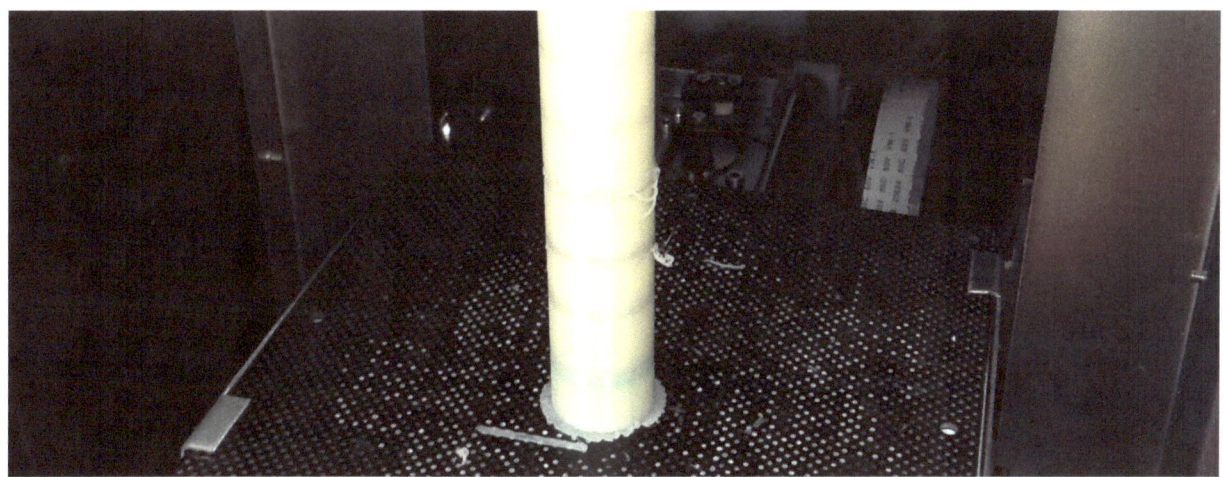

Rates & Ratios

WHAT YOU WILL NEED
- 3D printer
- White printer filament
- Cylinder .STL file to print
- Permanent markers (various colors)
- Tape measure (cm)

WHAT EACH STUDENT NEEDS
- Computer with SketchUp
- Ratio tables
- Proportions Practice Questions Handout
- Pencil

OVERVIEW & PURPOSE
In this lesson, students will use ratios and proportions to relate different variables within the 3D printing process, such as the volume of an object, the time it takes to print, and the amount of filament used in the print.

MATH COMMON CORE STANDARDS

CCSS.MATH.CONTENT.6.RP.A.1

Understand the concept of a ratio and use ratio language to describe a ratio relationship between two quantities.

CCSS.MATH.CONTENT.6.RP.A.3

Use ratio and rate reasoning to solve real-world and mathematical problems.

CCSS.MATH.CONTENT.7.RP.A.1

Compute unit rates associated with ratios of fractions, including ratios of lengths, areas and other quantities measured in like or different units.

CCSS.MATH.CONTENT.7.RP.A.2

Recognize and represent proportional relationships.

CCSS.MATH.CONTENT.8.EE.B.5

Compare two different proportional relationships represented in different ways.

A New Dimension of Mathematics with 3D Printing: **Grades 6 - 8**

Rates & Ratios

THE ACTIVITY
Lesson Duration: 50 minutes

Thinking and Sharing

For this lesson, 3D print with white filament so that markings can be made on it. Before class, use a Sharpie and tape measure to make marks on the filament every 100cm, to make measuring the amount used in the print easier.

For the purposes of this lesson, consider the starting point (0cm) to be the point where the filament enters the nozzle. This is not the same point as where the filament is extruded from the nozzle, but if kept consistent it will work for this lesson.

At the beginning of class, 3D print a cylinder with a radius of 1cm and a height of 7cm. Use the "solid" infill preference.
For later reference, the volume of this cylinder is given by the formula on the left. Either provide the students with this information, or have them calculate it using the formula for the volume of a cylinder if they are already aware of this formula.

Start the timer as soon as the printing begins. This print should take about 45 minutes. Tell the students they will use this cylinder print to explore ratios and proportions. Have the students observe the cylinder printing up close. This can be done by having students view the 3D printer a few at a time, or by attaching a webcam to the inside of the 3D printer and projecting a live video for all the students

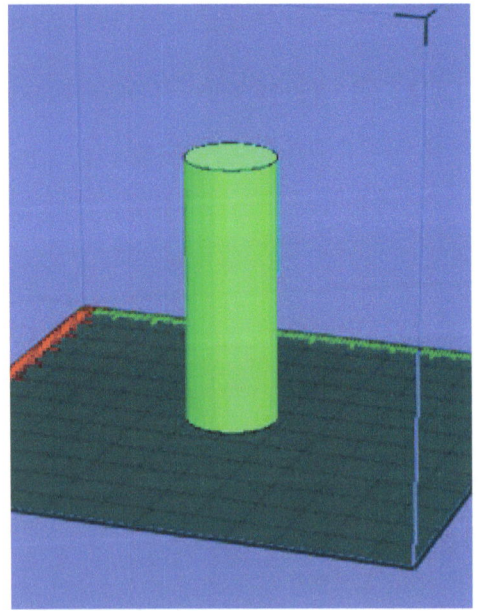

Cylinder with a radius of 1cm and a height of 7cm

A New Dimension of Mathematics with 3D Printing: **Grades 6 - 8**

Rates & Ratios

$$V = \pi(r^2) \cdot 7 = 7\pi \approx 22 cm^3$$

Formula for volume of the 3D printed cylinder.

to watch. Note that as layers are added and the volume of the shape increases, the printer uses more and more filament. So the volume of the shape and the amount of filament used have a directly proportional relationship. Additionally, as more layers are added and the volume of the shape increases, more time is needed to print the shape. So the volume of the shape and the amount of time it takes to print the shape are directly proportional as well.

Have students return to their seats. Emphasize that ratios and proportional relationships are very important in 3D design and 3D printing. There are many variables within the 3D design and printing process that are related to each other. As the students observed before, an object with a large amount of volume requires more time and filament to print. Additionally, a larger object would take more money to print. There are also differing qualities of 3D prints. For example, the 3D printing software allows the user to choose different infill options for the print, such as "solid," "hollow," or even "catfill."

Show students a picture of some of these different infills, and ask which one they think will use more filament when it prints. Which one do they think will print faster? (Answer: A more solid infill will use more filament and take more time. A more hollow infill will use less filament and take less time.) These questions allow students to see the ratios and proportions that are a natural part of 3D printing.

For this lesson's activity, the students will be exploring some of these proportional relationships using the

A New Dimension of Mathematics with 3D Printing: **Grades 6 - 8**

Rates & Ratios

Examples of three different infill options

cylinder that is currently printing. Hand out two tables to each student for them to record data. (possible picture of tables) The first table compares the volume of the cylinder with the amount of filament used to print the cylinder. The second table compares the volume of the cylinder and the time it takes to print the cylinder. Alternatively, all three of these variables could go into one chart rather than two; however, two separate charts may help reduce confusion when students do calculations later on.

Pause the printing cylinder and the timer at this point. The students will now use the information about the print so far to fill in the first row of both tables. First, the amount of time the print has taken so far is given by the timer. Second, calculate the current volume of the cylinder. The 3D printing software can show how many layers of the object have been printed so far out of the total number of layers. Use this information as well as the total volume of the shape to set up a proportion, which can be used to calculate the current volume of the solid. For example, if the print is currently on layer 76 out of 245 layers, and we know the total volume of the object will be about $22 cm^3$, the proportion shown to the left can be used to solve for the current volume of the object (x).

Table 1

Object	Volume of Shape (cm^3)	Amount of Filament Used (cm)
Partial cylinder		
Full cylinder		

Table 2

Object	Volume of Shape (cm^3)	Time to Print (minutes)
Partial cylinder		
Full cylinder		

Students should record data on tables.

Quickly calculate this current value and provide it to students to record in their table. Alternatively, students may calculate this value themselves with teacher guidance using their understanding of proportions. Different approaches to this problem are valid based on your students' exposure to proportions.

Finally, the students need to know how much filament

A New Dimension of Mathematics with 3D Printing: **Grades 6 - 8**

Rates & Ratios

$$\frac{76}{245} = \frac{x cm^3}{22 cm^3}$$

Use this proportion to solve for the current volume of the object.

has been used for the print so far. Have the students gather around the printer, and ask one or two students to examine the filament and determine how much has been used so far. They will do this by using the tape measure and the 100cm markings on the filament.

Start printing again so that the cylinder will finish printing, and continue timing the print with the timer. Have students return to their seats and make sure they have the first row filled in on both charts. Project a teacher or student copy of the tables with this information filled in so that students can catch up if they were behind or not paying attention. Ask the students: With the information that you have, do you think you could fill in the second row of the tables? That is, knowing the current volume, time, and amount of filament, could you determine the total volume, time, and amount of filament? This should generate some mathematical discussion, which can take place within groups and then as a class. The students should conclude that they need some further information. At this point, supply them with the total volume of the shape. So in the second row in the volume column, the students can fill in 22cm³, which is the total volume of the shape. Fill in this information on the projected chart.

On Their Own

Put students in teams of three or four, and tell them they are going to have a competition. Each group will have two minutes to fill in the blank in the first chart; that is, to determine how much filament will be used for the entire print, now that they know the final volume. At the end of the two minutes, the teams will provide their

Rates & Ratios

final estimation to the teacher to be recorded. When the cylinder is finished printing, the students will be excited to see which team was the closest!

It may be useful to have students think about the problem on their own first, and then compare answers with their group to come up with a collective answer. Have this stage be more student-focused, allowing the groups to come up with different answers even if they are wrong, which in turn encourages a development of their mathematical reasoning and sense making skills. Have teams write down their answer so other groups do not overhear and copy them. Once every group has an answer, have one person from each team select a permanent marker of a different color, marking their estimate directly on the filament using the tape measure and 100cm markings. Tell the students that when the print is finished, they will be able to see which team was the closest because the sharpie color will be printed along with the plastic filament. Throughout the remainder of the lesson, announce if a team's color is extruded, meaning their estimate was too small. When the print is finished, the final cylinder will have colored layers that show where each of those teams' estimates were. As students mark on the filament, ask: Now that you have seen how much filament that is, do you think your calculation is correct? Why/why not?

Now do the same thing with the second chart. Teams will have two minutes to determine how much time the total print will take. Once the print is finished, the timer will be checked to see which team was closest to the actual value. Go around and record each of the teams' estimations.

A New Dimension of Mathematics with 3D Printing: **Grades 6 - 8**

Rates & Ratios

Thinking and Sharing

At this point, go over how to use ratios and proportions to solve problems like this. This portion of the lesson may be short if the students only need to review this information, or it may be longer if this concept is new or . Go over some 3D printing examples for the class demonstrating how to set up and solve proportions. Alternatively, have students work out and explain these examples to the class. One example that could be worked out is shown below:

We can print an object with a volume of 8cm³ in 5 minutes. How long would it take to print an object with a volume of 30cm³?

To solve this problem, proportion on the left can be set up, since we can print 8cm³ every 5 minutes:

$$\frac{8cm^3}{5mins.} = \frac{30cm^3}{xmins}$$

We can then solve for x using a variety of methods to determine that it would take about 18.75 minutes.
Other examples could be similar, using relationships between the volume of an object, the amount of time it takes to print, and the amount of filament it takes to print.

On Their Own

Have students complete practice problems similar to the examples. Give each student a copy of the "Proportion Practice Problems" handout. The students may work with a partner or group. The problems require students to use the information they recorded in their tables earlier. If desired, the students may use the additional space in the tables to help work out or organize the solutions to the problems on the handout. Assist students as needed,

A New Dimension of Mathematics with 3D Printing: **Grades 6 - 8**

Rates & Ratios

encouraging students to justify their answer and explain what the answer means in the context of the problem. If all the students are stuck on a particular problem, work that problem out with the whole class. Students will likely not finish all the practice problems. These can be used for further practice during a later lesson or for homework.

Thinking and Sharing

Near the end of class, the cylinder will finish printing. Make sure to stop the timer as well. Have all the students gather around to see which colored marking is closest to the nozzle and one might even be in the nozzle. This team had the closest approximation to the real answer, so they are a winner! Determine how much time the print took, and see which team's time estimate was closest to the actual time. This team is also a winner! Ask questions such as:

- This team's mark on the filament has not made it near the nozzle yet. Was their guess too high or too low? (Answer: Too high)
- This team's color got pushed through the nozzle too early. Was their guess too high or too low? (Answer: Too low)

Next, actually calculate the amount of total filament that should have been used and the total amount of time that the print should have taken, based on the initial values recorded in the first row of the charts. Use proportions to find what these values should have been exactly. For example, set up the proportion on the left to find the total amount of filament that the print should have used:

A New Dimension of Mathematics with 3D Printing: Grades 6 - 8

Rates & Ratios

$$\frac{length\ of\ filament\ after\ pausing\ print}{volume\ of\ object\ after\ pausing\ print} = \frac{total\ length\ of\ filament}{total\ volume\ of\ object}$$

Use this proportion to find the total amount of filament used by the print.

Compare these calculated values to the actual final observed values for total filament and total time of the print. The calculated values may vary slightly from the actual observed values.

FOR THEIR PORTFOLIO

- Initial attempt to solve proportion problem (the group competition)
- Practice problems on worksheet

GOING FURTHER

- Proportions are related to linear relationships. Have students graph the linear relationship between two variables of the printing process, such as volume of an object and time to print.
- Add cost of a 3D print as another variable in this lesson. For example, tell the students how much each centimeter of filament costs (unit rate) or the cost and length of a spool, and have them determine how much it will cost to print an entire object. An additional option would be to tell students they want to print a certain object but do not want to go over a certain price. Ask them if they think they will be able to print the object. Alternatively, students could build an object out of unit cubes, imagining that each unit cube costs a certain amount of money, and then calculate the cost of printing their design.
- Students could use proportional reasoning and/or a table to compare the quality of a 3D-printed object and the speed at which the object prints. In general, the thicker the layers are, the poorer the quality will be of the resulting object, but the faster it will print.
- When the print of the cylinder is first paused, this can

Rates & Ratios

be a time to emphasize ideas of cross sections as well. For example, the students will notice that each layer building up the cylinder so far has been a circle. They may predict that they layers will continue to be circles until the shape has reached its full height. These circles are cross sections of the cylinder horizontal to the base.

TEACHER TALK

When the students first estimate the total time and filament in their teams, it is likely that different students will use a variety of approaches to tackle the problem. Some may simply estimate the total time or amount of filament using the current information. Some may approximate proportional relationships or use informal proportional reasoning. Some may remember and use more formal methods for finding missing values using proportions. All of these approaches are valuable.

If students have not had very much formal instruction related to proportional reasoning, this lesson may need to be more intuitive and estimation-based, rather than using proportions and formal methods of solving them.

TECHNOLOGY TIPS

- To pause the print, go to 3D print -> Maintenance and then select the Pause option
- Sometimes the 3D printing software does not properly restart the print after it has been paused. It may be wise to plan a backup option in case this happens.

Rates & Ratios

OBSERVATION TABLES

Name _____

Record your findings here:

Object	Volume of shape (cm³)	Amount of filament used (cm)
Partial cylinder		
Full cylinder		

Object	Volume of shape (cm³)	Time to print (minutes)
Partial cylinder		
Full cylinder		

A New Dimension of Mathematics with 3D Printing: **Grades 6 - 8**

Rates & Ratios

PROPORTION PRACTICE   Name _____

Use the information in your tables to answer these questions:

1. Circle the object that would print the fastest.

Volume=255cm^3

Volume=304cm^3

2. Circle the object that would use the most filament.

Volume=255cm^3

Volume=304cm^3

3. The volume of this object is 50cm^3

a. How long would it take to print this soccer ball?
_____minutes

b. How much filament would we need to print this soccer ball? Show your work.
_____minutes

A New Dimension of Mathematics with 3D Printing: **Grades 6 - 8**

Rates & Ratios

PROPORTION PRACTICE

Name _____

4. This castle took 25 minutes to print. What is its volume?

5. It took 50cm of filament to print Mario. What is the volume of this object?

6. If an object takes 60 minutes to print, how much filament is being used?

7. In class we printed one cylinder with volume 20cm^3. How long would it take to print 5 of these cylinders?

8. How many of these cylinders could you print in a day (24 hours)?

A New Dimension of Mathematics with 3D Printing: **Grades 6 - 8**

Cookie Cutter Tessellations

WHAT YOU WILL NEED
- Printed tessellation cookie cutter
- 3D printer
- Method of saving student files

WHAT EACH STUDENT NEEDS
- Computer with SketchUp
- Tessellation SketchUp guidelines
- Tessellations follow-up handout
- Exit slip questions
- Blank paper

FOLLOW UP MATERIALS
- Students' printed cookie cutters
- Oven (can be a convection oven)
- Oven mitt
- Cookie sheets & Wax paper
- Cookie dough
- Spatula
- Fridge (for cookie dough)

OVERVIEW & PURPOSE
Students will create 3D tessellation designs that can be used as cookie cutters. Students can test the tessellation designs by determining if their cookie cutter can cut out cookie dough without any gaps or overlaps. The lesson also explores which regular polygons tessellate a plane and which others do not.

MATH COMMON CORE STANDARDS
CCSS.MATH.CONTENT.7.G.B.5

Use facts about supplementary, complementary, vertical, and adjacent angles in a multi-step problem to write and solve simple equations for an unknown angle in a figure.

CCSS.MATH.CONTENT.8.G.A.1

Verify experimentally the properties of rotations, reflections, and translations.

CCSS.MATH.CONTENT.8.G.A.2

Understand that a two-dimensional figure is congruent to another if the second can be obtained from the first by a sequence of rotations, reflections, and translations; given **two congruent figures, describe a sequence that exhibits the congruence between them.**

A New Dimension of Mathematics with 3D Printing: **Grades 6 - 8**

Cookie Cutter Tessellations

Honeycombs are a great example of tessellations in nature.

THE ACTIVITY

Lesson Duration: 50 minutes (Note: This lesson requires a 10-20 minute follow-up in a later lesson)

Thinking and Sharing

This activity has students design and 3D print a usable cookie cutter using translations and rotations. Since this cookie cutter will be a tessellation, make sure to ask students if they have ever seen or heard this term before. Inform the students that a tessellation is a continuous repeated pattern of shapes that covers the plane with no gaps or overlaps. If students are unfamiliar with the term "plane," they can imagine a plane as the floor or a piece of paper, that goes on forever. Ask students to list some examples of tessellations they have seen in real life. If they cannot come up with any, have them look around the classroom for ideas. Some common examples are floor tiles, brick walls, puzzle pieces, checkerboards, and honeycomb. Tessellations are also used by some artists. Show students some pictures of these tessellations in real life, as seen in the picture on the left.

Additionally, show the students some pictures of non-examples of tessellations, and ask them to explain why they are not tessellations. The students should recognize that these shapes overlap or leave gaps, so they are unable to perfectly cover a plane. Some examples are shown on the left.

On Their Own

Ask students to explore which regular polygons can tessellate a plane. Remind the students that a polygon is a

A New Dimension of Mathematics with 3D Printing: **Grades 6 - 8**

Cookie Cutter Tessellations

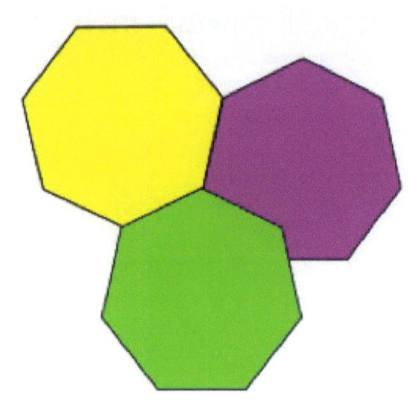

Shapes that overlap or leave gaps cannot tessellate a plane.

2D shape with straight sides, such as a triangle, rectangle, or pentagon. A regular polygon is a polygon where all the sides and all the angle measurements are the same, such as an equilateral triangle or a square. First, have students think on their own about which regular polygons can tessellate a plane. If desired, they may use SketchUp and the Shape tool to create and manipulate polygons. This may help the students visualize the shapes coming together at a corner. Have students write down their ideas. After a few minutes, the students will compare their thoughts with a partner.

Thinking and Sharing

Bring the class back together. Show the class pictures of all the regular polygons up to the decagon, such as the one shown on the left.

With each picture above, ask the class if they think that shape will tessellate a plane. Have students explain why they think that shape will or will not tessellate.

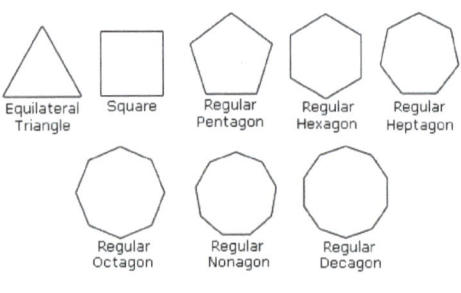

All regular polygons up to the decagon.

As necessary, help clarify and expound students' reasoning about which regular polygons will tessellate the plane. One such line of logic might start by drawing a point on the board, asking students to imagine they are going to walk all the way around it. By proposing this observation, students should come to the conclusion that they will walk in a full circle, or in 360 degrees. Then continue this example by drawing a square, so that a single corner of this square touches the point previously drawn on the board. Students should note that the interior angles of the square are all 90 degrees. Draw one more square, connecting it to the edge of the first square as well

A New Dimension of Mathematics with 3D Printing: **Grades 6 - 8**

Cookie Cutter Tessellations

as placing one corner on the central original point.

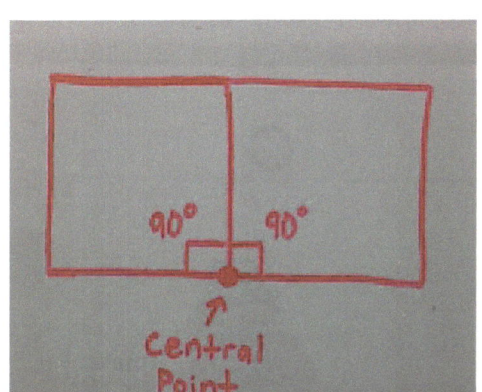
Showing degrees with squares

The interior angles of this square are also 90 degrees. At this point, the angles that the squares have created at the original point add up to 180 degrees. Ask the students how many squares it will take to get to 360 degrees at the point they all touch. The students should conclude that the number of squares is four, which can be clarified by the equation 4*90°=360°. Four squares, therefore, will all meet up perfectly at a vertex, with no little gaps or overlaps. We can repeat this pattern of squares continuously around every vertex. Therefore a square will tessellate the plane. This can be extended to rectangles as well, because rectangles also have four angles that are each 90 degrees.

Perform this same demonstration with an equilateral triangle. Start with a point and draw triangles all around that point. These triangles have interior angles of 60 degrees each. It should take six triangles to go around the point, because 6*60°=360°. Then perform the same experiment with a hexagon, pentagon, and heptagon. As an alternative to drawing these on the board, one option is to split the class into three groups and have one group try the hexagon, one group try the pentagon, and one group try the heptagon. Students should conclude that the hexagon, with interior angles of 120 degree, will tessellate a plane because 3*120°=360°. However, a pentagon has interior angles of 108 degrees. There is no integer we can multiply by 108° to get an even 360 total degrees around a vertex. Students will notice that after they draw three pentagons connected at a single vertex, there is a small gap left between two of the pentagons, meaning the pentagon cannot tessellate the plane. Heptagons have an

A New Dimension of Mathematics with 3D Printing: **Grades 6 - 8**

Cookie Cutter Tessellations

interior angle of approximately 129 degrees. If students draw three heptagons that connect at a single vertex, they will note that an overlap is created. Thus, the interior angles of heptagons are too wide, meaning it cannot tessellate the plane. As an extension, any other regular polygon with more sides than a heptagon will also have interior angles that are too wide; therefore, these other polygons are not tessellations. If the students desire, they can try out these other polygons later on their own. In conclusion, triangles, squares, and hexagons are the only regular polygons that can tessellate a plane.

Next, begin the cookie cutter activity. Ask students if they have ever made cookies by rolling out the dough and cutting the dough with cookie cutters. Often, when people cut cookie dough in this way, they try to fit each cut with the cookie cutters as close together as possible, so that there is minimal leftover dough that will have to be rolled out again. The cookie cutters that the students design today will be tessellations, so when they are used to cut the dough, there will be absolutely no overlaps or leftover dough (except on the edges).

The students will design their tessellation by starting with a rectangle or square. The students know that fitting rectangles together will tessellate a plane, so a rectangular cookie cutter could perfectly cut out the dough with no gaps or overlaps. However, ask the students what they think would happen if they cut out part of the rectangle, from one vertex to an adjacent vertex, and translate (or slide) this piece from one edge of the rectangle to the other. Would these new shapes still fit together perfectly? What if we did this to two edges? These shapes should

A New Dimension of Mathematics with 3D Printing: **Grades 6 - 8**

Cookie Cutter Tessellations

still tessellate, because the shapes will still fit together perfectly. The translation preserved the congruency of the shape. This is similar to the way puzzle pieces are created.

On Their Own

Have students open SketchUp on their computers. Also, pass out the "Guidelines for Creating a Tessellation" handout to each student. Using this handout as a reference, demonstrate for the students how to create a tessellation in SketchUp. In this handout, there are step-by-step directions for the teacher's use as well as student use. Either have the students watch this first demonstration, or have them follow along with each step on their own computers. The students may also pass around a demo cookie cutter that the teacher printed before class, and observe the teacher's SketchUp design that goes along with the cookie cutter. This will give the students a better idea of what to design.

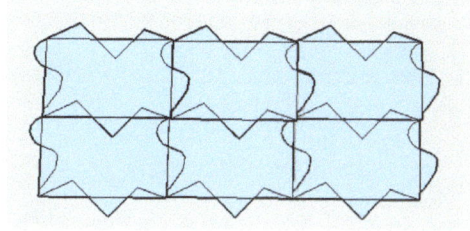

Rectangles to puzzle pieces- pictures in lesson on Drive

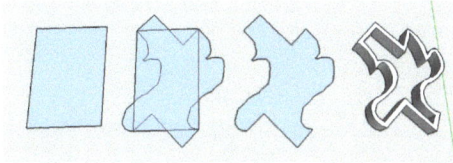

After this, instruct the students to open a new SketchUp document, and to begin designing their own tessellation. Students can actively check to see if their design tessellates the plane by making copies of the shape and fitting them together. If time allows, have the students create three tessellation designs and pick their favorite to be printed. Advanced students could create even more designs, or they could experiment with tessellations beginning with a triangle or hexagon rather than a rectangle. With a few minutes left in class, each student should pick their favorite tessellation and save it to be 3D printed as a cookie cutter! Ask students: Do you think your design will cut out cookie dough with no gaps or

A New Dimension of Mathematics with 3D Printing: **Grades 6 - 8**

Cookie Cutter Tessellations

overlaps? They will test these cookie cutters out on a later day.

Thinking and Sharing

Conclude the lesson by summarizing the students' discoveries about tessellations, the regular polygons will tessellate the plane, and how those shapes can still tessellate even when modified. Have the students complete a quick exit slip, such as the "Tessellations Exit Slip" handout.

After Class

After class, open all of the students' cookie cutter designs in SketchUp. These designs should currently be two-dimensional polygons. These designs need to be turned into a 3D printable format that will cut cookie dough. First, the inside of the design needs to be eliminated so that it can actually cut cookie dough. To do this, open a design and create a copy of the design that is slightly smaller in scale. This can be done by manually creating new lines along the inside of the shape that mimic the current lines created by the student. The inside of the shape can then be deleted so that all that is left of the design is an outer outline of the cookie cutter that will become the "wall." Try to make this area as thin as possible, but keep in mind that particularly thin walls might break. The design can now be turned into a 3D object using the Push/Pull tool, which will give the walls of the design height. Alternatively, give the students directions at the end of class to do this process themselves.

Part of the cookie cutter design process is shown in the picture on the left.

A New Dimension of Mathematics with 3D Printing: **Grades 6 - 8**

Cookie Cutter Tessellations

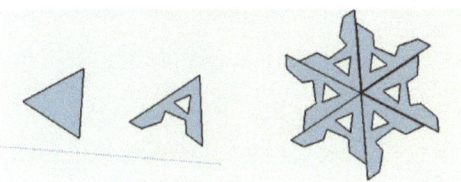

The final design in this process can be 3D printed.

Follow Up

In a later lesson, hand back the students' 3D printed cookie cutters for them to keep. Also hand out the "Tessellation Follow-Up" handout to each student. On the blank space on the back of the handout, students will trace around their design multiple times to see if it can tessellate the page. After this initial test, hand each student a small amount of cookie dough on wax paper for them to flatten. Then the students can use their cookie cutter to cut out cookies with their dough. Instruct the students to try to optimize the dough without having to roll it out more than once. Students should try to fit the cuts together perfectly, with no gaps or overlaps, proving that their design is a tessellation. If there is an oven at the school or a convection oven near the classroom, put the students' cookies on a cookie sheet and bake them. While the cookies are baking, the students can discuss their conclusions about tessellations from this entire activity, or you may combine this lesson with another lesson. Once the cookies are done baking, the students can eat them! Be aware that the cookies will expand in the oven, so the baked cookies will probably not tessellate as well as the dough.

Alternatively, if baking cookies at school is not an option, bake the cookies at home and bring them in for the students to enjoy, or the students could cut out shapes in playdough or clay rather than cookie dough. Another alternative is to make the designs into stamps rather than cookie cutters by leaving the designs solid and not taking out the middle of the shape.

A New Dimension of Mathematics with 3D Printing: **Grades 6 - 8**

Cookie Cutter Tessellations

FOR THEIR PORTFOLIO

- Students will design at least one cookie cutter that can be 3D printed and will tessellate a plane, as demonstrated by the students' justifications on the follow-up worksheet, and by the students' tracing around their design with a pencil.
- The students may complete an exit slip after the initial activity to assess their knowledge of tessellations and some of the related vocabulary, such as "interior angle."

GOING FURTHER

- Instead of translating the cut portion of the rectangle from one edge to another, advanced students could experiment with rotating these cut portions to another edge, and see if these new shapes will still tessellate.
- Tessellations can also be related to art topics. M. C. Escher is a popular artist with many works involving tessellations.
- For a paper version of this tessellation activity, see this website:

https://www.exploratorium.edu/geometryplayground/resources.php

- For an additional project with transformations, have students create a snowflake or sun in SketchUp that can be 3D printed using rotations and angle measurements. Students will create a triangle in SketchUp. They will then cut out portions of the triangle using the Pencil, Shape and Eraser tools. They will then make the triangle a group. This triangle can be rotated around one of its vertices six times, using

A New Dimension of Mathematics with 3D Printing: **Grades 6 - 8**

Cookie Cutter Tessellations

the copy and paste options and the Rotate tool. Since the interior angles of the triangle are each 60 degrees, the triangle must be rotated 60 degrees each time a new copy is made. After six copies of the triangle have been created, this should complete the 360 degrees around the point, and all the triangles will be touching at that central vertex. The students can now "explode" the shapes so that they are no longer groups. Then they will use the Eraser tool to get rid of unwanted lines within the shape, and use the Push/Pull tool to make the design 3D. Students can make the design all the same height, or they could make certain parts different heights. More advanced students could repeat this activity starting with a square or hexagon rather than a triangle. At the end of this activity, have students make conclusions about interior angles of shapes, as well as the congruency of each copy of the initial triangle. Some examples are shown on the left.

TEACHER TALK

Tessellations are valuable extensions of transformations, particularly translations and rotations, as well as knowledge of angles, such as interior angles of a regular polygon. Also, this lesson allows students to explore tessellations first through angle measurements and mathematical calculations. Then they are able to further explore tessellations on their own using the SketchUp software, and eventually with a physical printed copy of their tessellation design. By physically cutting out the cookie dough so that all the cuts fit together perfectly, the students are provided with a strong visual to remember the definition of a tessellation, as well as how translations preserve congruence.

A New Dimension of Mathematics with 3D Printing: **Grades 6 - 8**

Cookie Cutter Tessellations

TECHNOLOGY TIPS

- As students are creating lines on their rectangle, they may occasionally need to orbit around the shape to make sure all their lines are remaining on the same plane and not stretching into another dimension. As mentioned on the "Tessellation Guidelines" handout, it helps to start by selecting the Top Standard View in the Camera menu to look straight down at the rectangle.
- When using the Eraser tool to eliminate the edges of original shape after the translations have been made (Step 7 on the "Tessellation Guidelines" handout), the inside of the shape may disappear. This is because the lines were not completely connected to form a complete polygon. If this happens, zoom in and create little lines to make the shape whole before deleting the edge.
- This activity works best when the students keep their design 2D while they are cutting and pasting pieces, and then make the shape 3D at the very end. It is possible for the students to start with a 3D shape, but they may need to add some extra lines to the sides in order to complete the design. This approach is definitely more difficult that the approach described in this activity.

Cookie Cutter Tessellations

CONSTRUCTION GUIDELINES

Name _____

1. Choose Top from Standard Views in the Camera menu so you are looking straight down at the workspace.

2. Create a rectangle.

3. Using the Pencil tool, draw several connected line segments along the top line, starting at the left corner and ending at the right.

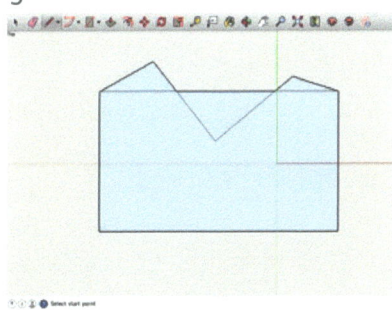

4. Use the black Arrow tool to select all the triangular shapes you just created. Do this by Shift-clicking on each shape.

5. Next, click on the Move tool (red arrows) and move the mouse to the top left endpoint of the rectangle. Press the Option key, then click and drag the mouse down to the lower left endpoint. This moves a copy of the selected triangles to the bottom of the rectangle.

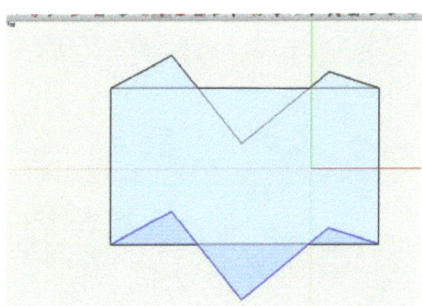

A New Dimension of Mathematics with 3D Printing: **Grades 6 - 8**

Cookie Cutter Tessellations

CONSTRUCTION GUIDELINES

Name _____

6. Repeat this process with other shapes on the right edge of the rectangle. Select it and move a copy to the left edge.

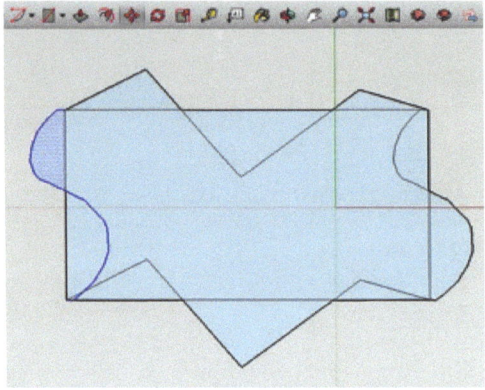

7. Use the eraser to erase all original rectangle lines.

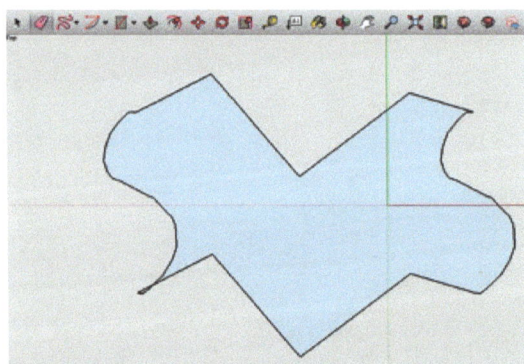

8. You can copy and paste your shape to make sure it tessellates!

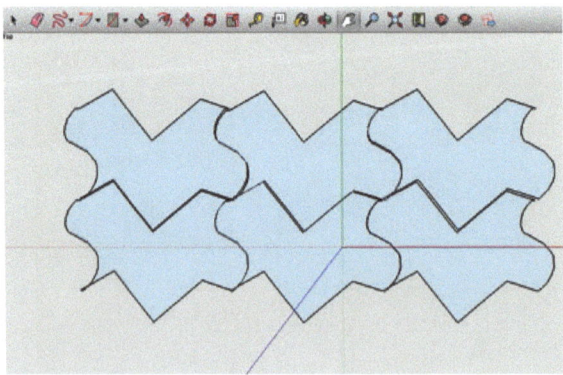

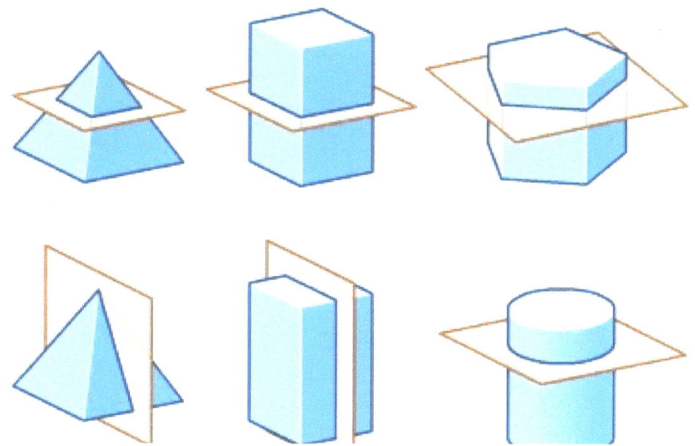

Cross Sections

WHAT YOU WILL NEED

- 3D printer
- Method of saving student files
- Time-lapse recording of the six shapes being printed
- Multiple copies of six pre-printed shapes (triangular prisms, rectangular prisms, cylinders)
- Pre-printed shapes that were stopped before printing completed
- .STL file of six solids
- A video of any object printing
-

WHAT EACH STUDENT NEEDS

- Computer with SketchUp
- Review activity handout
- Partner activity handout
- Pencil

OVERVIEW & PURPOSE

In this lesson, students will compare and contrast the cross sections of triangular prisms, rectangular prisms, and cylinders. They will also explore multiple ways to construct the same geometric solid.

MATH COMMON CORE STANDARDS

CCSS.MATH.CONTENT.7.G.A.3

Describe the two-dimensional figures that result from slicing three-dimensional figures, as in plane sections of right rectangular prisms and right rectangular pyramids.

CCSS.MATH.CONTENT.8.G.A.1

Verify experimentally the properties of rotations, reflections, and translations.

A New Dimension of Mathematics with 3D Printing: **Grades 6 - 8**

Cross Sections

THE ACTIVITY
Lesson Duration: 50 minutes

Thinking and Sharing

Note that this lesson has two components: observing the cross sections of geometric solids and observing the congruence of two geometric solids that are rotations of each other. Alternatively, these could be split into two separate lessons, or spread over two days.

Start the lesson by reviewing how the 3D printer actually works. In particular, draw connections between the Push/Pull tool in SketchUp and the way the 3D printer extrudes plastic. In SketchUp, the Push/Pull tool is how we turn a 2D solid into a 3D object. When we send a 3D design to the 3D printer, the printer knows to split the design into many tiny, horizontal layers. The printer extrudes the bottom layer onto the print bed, then the next layer, then the next, continuing in this pattern until it has built the entire object.

This lesson will focus on the exploration in cross sections, which are 2D shapes that result from cutting 3D shapes with a plane. Depending on how we cut through a solid, the cross sections may be different. When the 3D printer extrudes one layer of a solid at a time, students can imagine those layers as cross sections. That is, each little layer of the 3D object is one of its cross sections.

Let students pass around some 3D printed shapes that were stopped in the middle of the printing process before they were finished printing. Students will observe the

Cross Sections

cross sections of the shapes at the moment when they stopped printing, concentrating on the top-most layer. If printed objects are not available, the students can simply observe a picture of such objects or watch a video online of 3D objects printing. Stop the video presentation in the middle of the printing process at certain intervals to show the students the solid of the current layer being printed, making note that the individual layers are cross sections of the solid. Some questions to ask the class are:

- Is a cross section 2D or 3D?
- Do all the cross sections for a solid have to be the same? Why/why not?
- What do you predict the next cross section will look like?

For the rest of the lesson, students will be working with triangular prisms, rectangular prisms, and cylinders. If the students need a more in-depth review of these terms, have them complete the "Review Activity" worksheet individually or in pairs.

On Their Own

For this activity, split the class into three groups, assigning each group a solid: rectangular prism, triangular prism, or cylinder. A differentiation option is to place less advanced students in the rectangular prism group, since this is the most simple solid. Within each group, have each student work with a partner. The students are going to design their group's solid in SketchUp; however, each pair will be given two handouts that depict the same solid in different orientations (see "Cross Sections" handout). One partner should be given the handout where the solid

Cross Sections

is standing on its base, and the other partner should be given the handout where the solid is on its side. There are six different handouts, which includes two versions for each of the three shapes.

These handouts will lead one partner to design the group's solid with the base on the bottom. The other partner will design the solid with the base on the side. The students may consider this orientation "sideways" if they are used to seeing the solid with the base on the bottom. For example, the first partner would design a triangular prism standing upright with the triangular base on the bottom. The second partner would create a triangular prism where one of the rectangular sides is on the bottom and the two triangular bases are on the ends of the solid. There are specific measurements included in the design directions on the handout, so students' designs should be exactly the same.

Simply hand the students the directions for constructing their particular solid in SketchUp. Do not explain to the students that their solid should be congruent to their partners' solid. This activity works best if the students discover on their own that the two shapes have the same measurements. Once they have finished their design, the students will compare their solid with their partner's solid. They should observe that their solid is congruent to their partner's solid. Encourage students to use the orbit tool to rotate around their solid until it looks like their partner's solid. Their solid is simply a rotation of their partner's solid. Student observations will be guided by the handout, which asks students to respond to these questions:

A New Dimension of Mathematics with 3D Printing: **Grades 6 - 8**

Cross Sections

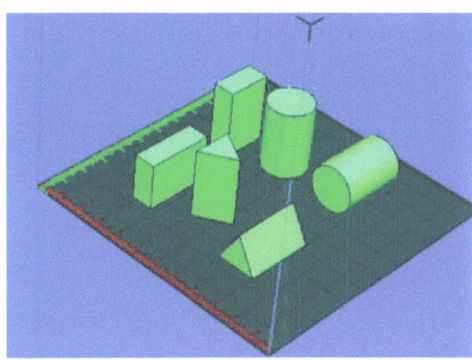
Six solids on a print bed

- Compare your solid with your partner's solid.
- What will the cross sections of your solid look like when you cut parallel to the base?
- What will the cross sections of your solid look like when you cut vertically through your solid?
- When you 3D print your object, what solid will the layers be that are being printed?

These are additional questions that could be asked as students design the shapes:

- How do you know your solid is the same as your partner's solid?
- How is a horizontal cross section related to the layers extruded by the 3D printer?

Thinking and Sharing

Have the students share the results that they found in the previous activity to the class. This can be done as a whole class, or one student from each group can form new groups of three to share their results.

Next, 3D print the six shapes that the students designed all at the same time. These six shapes will all be printed on the print bed together as in the picture on the left.

Cross Sections of six solids on a print bed

Have a time-lapse video of this same print pre-recorded to show the class, so all the students can observe it quickly and at the same time. As students watch the video, have them observe the layers being printed pausing the video occasionally to discuss. Ask the students:

- Which solid do you think is the one you designed?
- Your partner?

A New Dimension of Mathematics with 3D Printing: **Grades 6 - 8**

Cross Sections

At first, a few of the solids will immediately stand out. For example, the upright cylinder is easy to identify, because it is the only one with a horizontal cross section that is a circle. Some of the solids will be more difficult to distinguish. For instance, the "sideways" triangular prism and the "sideways" rectangular prism will be printed very similarly at first. Both will begin with a rectangular prism. However, as more layers are added to the solids, each rectangular cross section of the triangular prism gets smaller while the rectangular layers of the rectangular prism stay the same size throughout the print. Eventually, allow the students to gather around the 3D printer a few at a time to observe the actual printing process taking place. The same questions from before can be asked again during this time.

Next, the students will physically examine these six geometric solids, including two printing orientations for each of the three solids. For this part of the lesson, make sure to have prepared and 3D printed these six solids before class since there will not be enough time to print the students' designs in class. These solids, therefore, should be identical to the ones the students just designed. If possible, print multiple copies of each solid so that every student can have one of the six solids. Ideally, each student should be given a physical copy of the solid they just designed. As students observe these physical shapes, have them compare the solid they designed with the shapes designed by their partner, their group, and the class.

Cross Sections

Start by asking the students to compare their solid with their partner's solid. The students should be able to clearly see that their solid and their partner's solid are the same even though they were designed differently. Also, the students should be able to observe tiny striations on their printed shapes that reveal the different layers from the printing process. When comparing their shapes with their partner, students should observe that their partner's solid has layers that are different than their own, because the two shapes were printed in different orientations. After these observations have been made, have students examine shapes from other groups to see how the same concepts can be applied to the other shapes.

Alternatively, the students could actually 3D print each of their designs rather than using pre-printed shapes. Then these observations and comparisons between printed shapes could take place during a later lesson.

Conclude by reemphasizing the students' two main discoveries from the lesson:
1. 3D shapes do not have to be restricted to always having the typical base on the bottom, such as a circle for a cylinder or a triangle for a triangular prism. Solids designed in other orientations ("sideways" shapes) are congruent to regular shapes of the same measurements. The "sideways" solid is simply a rotation of the original solid.
2. Cross sections are 2D slices of a 3D solid. In particular, cross sections horizontal to the base of a solid can be pictured as the layers that the 3D printer extrudes as it builds up an object. Cross sections can help to distinguish one geometric solid from another.

A New Dimension of Mathematics with 3D Printing: **Grades 6 - 8**

Cross Sections

FOR THEIR PORTFOLIO

- The student SketchUp designs of the six solids assigned to them should be saved.
- The students' observations on the worksheet comparing their solid with their partners' shapes.

GOING FURTHER

- As an extension, have students explore the volume of oblique objects versus the volume of right objects with the same base and height. Students can observe in the context of 3D printing that two 3D objects have the same volume so long as every single cross section of the shapes is the same, regardless of whether the cross sections are skewed or not. This is easily pictured using the layers of a 3D print.
- To challenge more advanced students to think about real world design, ask: Which way do you think is the best way to orient your solid when you're printing? There is not necessarily a right answer to this question, but students will have to extend the idea of the layers being printed on top of each other.

TEACHER TALK

- 3D printing is a powerful teaching tool when introducing cross sections, because the layers extruded during the printing process allow students to visualize the shape of each horizontal cross sections. For instance, students might not be able to understand why a cylinder starts as a circle, but by watching the object being created layer by layer the student is physically able to see the height of the object being built up before their eyes, one circular

Cross Sections

cross section at a time.
- With 3D printing, students are not restricted to the orientation of shapes presented in a textbook or on a worksheet. They can rotate their shapes around in the 3D design software and then by holding the physical 3D printed objects in their hands. This emphasizes the idea that the base of a geometric solid does not necessarily have to be on the bottom of the solid, which is a common misconception for students.

TECHNOLOGY TIPS

- Students who are instructed to create their solid "sideways" may struggle to get their initial 2D base to be standing on end rather than flat. A good way to do this is to create a rectangular prism first. Then the faces of this prism and the Shift key can be used to orient other shapes the designer wants to make. For example, if the designer wants to make the circular base of a cylinder, they would select the Circle tool. Then they would put the cursor over the face of the rectangular prism that they want to be parallel to their circle. Then hold down the Shift key, which will orient the circle to be parallel to that particular face of the prism. While holding down the Shift key, move the cursor to the desired location and create a circle as normal.

Cross Sections

SHAPE REVIEW WORKSHEET   **Name** _____

1. Create a rectangle with the Shape tool. Use the Push/Pull tool to make your rectangle 3D.

   a. What shape are all the faces of your object? (Hint: Use the Orbit tool!)

   b. What is this object called?

2. Create a triangle with the Polygon tool. Use the Push/Pull tool to make your triangle 3D.

   a. What shape are all the faces of your object?

   b. What is this object called?

3. Create a circle with the Shape tool. Use the Push/Pull tool to make your circle 3D.

   a. What shape are the top and bottom of your object?

   b. What is this object called?

A New Dimension of Mathematics with 3D Printing: **Grades 6 - 8**

Cross Sections

PARTNER ACTIVITY WORKSHEET

Name _____

Construct this **triangular prism** in SketchUp. Measurements are in mm.

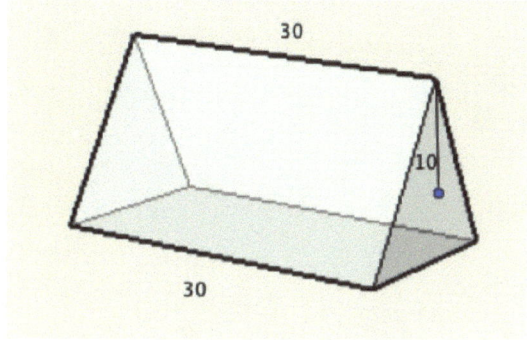

1. Compare your shape with your partner's shape. Be as specific as possible.

2. What will the cross sections of your solid look like when you cut parallel to the base?

3. What will the cross sections of your shape look like when you cut vertically through your solid?

4. When you 3D print your object, what shape will the layers be?

A New Dimension of Mathematics with 3D Printing: **Grades 6 - 8**

Cross Sections

PARTNER ACTIVITY WORKSHEET

Name _____

Construct this **triangular prism** in SketchUp. Measurements are in mm.

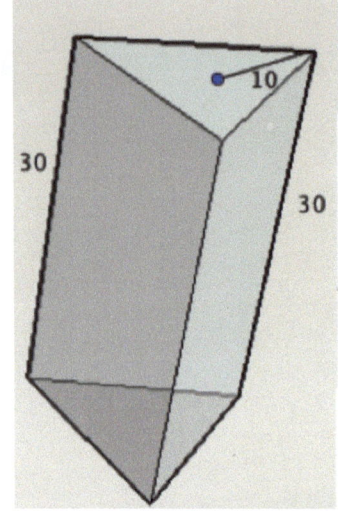

1. Compare your shape with your partner's shape. Be as specific as possible.

2. What will the cross sections of your solid look like when you cut parallel to the base?

3. What will the cross sections of your shape look like when you cut vertically through your solid?

4. When you 3D print your object, what shape will the layers be?

A New Dimension of Mathematics with 3D Printing: Grades 6 - 8

Cross Sections

PARTNER ACTIVITY WORKSHEET

Name _____

Construct this **rectangular prism** in SketchUp. Measurements are in mm.

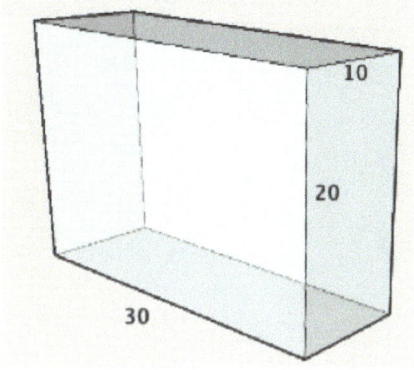

1. Compare your shape with your partner's shape. Be as specific as possible.

2. What will the cross sections of your solid look like when you cut parallel to the base?

3. What will the cross sections of your shape look like when you cut vertically through your solid?

4. When you 3D print your object, what shape will the layers be?

A New Dimension of Mathematics with 3D Printing: **Grades 6 - 8**

Cross Sections

PARTNER ACTIVITY WORKSHEET   Name _____

Construct this **rectangular prism** in SketchUp. Measurements are in mm.

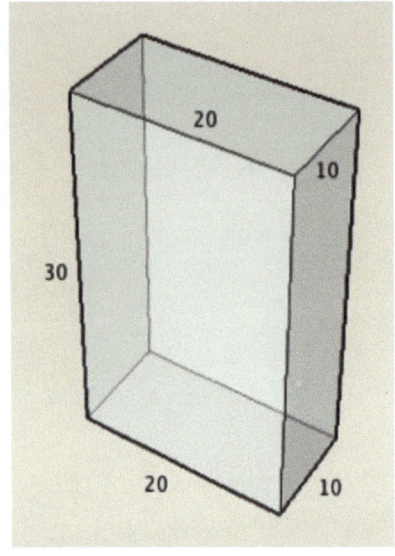

1. Compare your shape with your partner's shape. Be as specific as possible.

2. What will the cross sections of your solid look like when you cut parallel to the base?

3. What will the cross sections of your shape look like when you cut vertically through your solid?

4. When you 3D print your object, what shape will the layers be?

A New Dimension of Mathematics with 3D Printing: **Grades 6 - 8**

Cross Sections

PARTNER ACTIVITY WORKSHEET

Name _____

Construct this **cylinder** in SketchUp. Measurements are in mm.

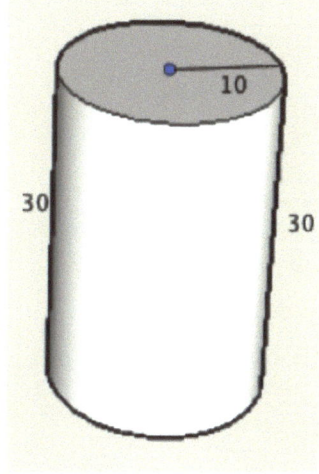

1. Compare your shape with your partner's shape. Be as specific as possible.

2. What will the cross sections of your solid look like when you cut parallel to the base?

3. What will the cross sections of your shape look like when you cut vertically through your solid?

4. When you 3D print your object, what shape will the layers be?

Cross Sections

PARTNER ACTIVITY WORKSHEET   Name _____

Construct this **cylinder** in SketchUp. Measurements are in mm.

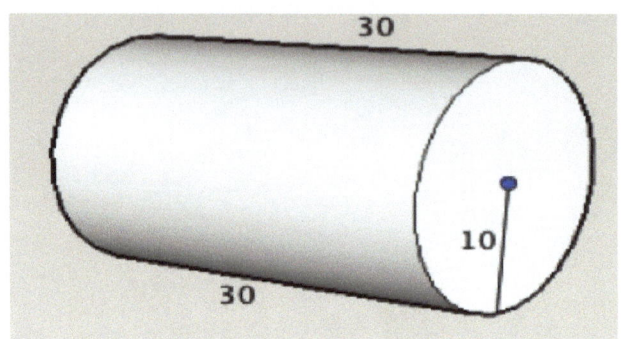

1. Compare your shape with your partner's shape. Be as specific as possible.

2. What will the cross sections of your solid look like when you cut parallel to the base?

3. What will the cross sections of your shape look like when you cut vertically through your solid?

4. When you 3D print your object, what shape will the layers be?

A New Dimension of Mathematics with 3D Printing: **Grades 6 - 8**

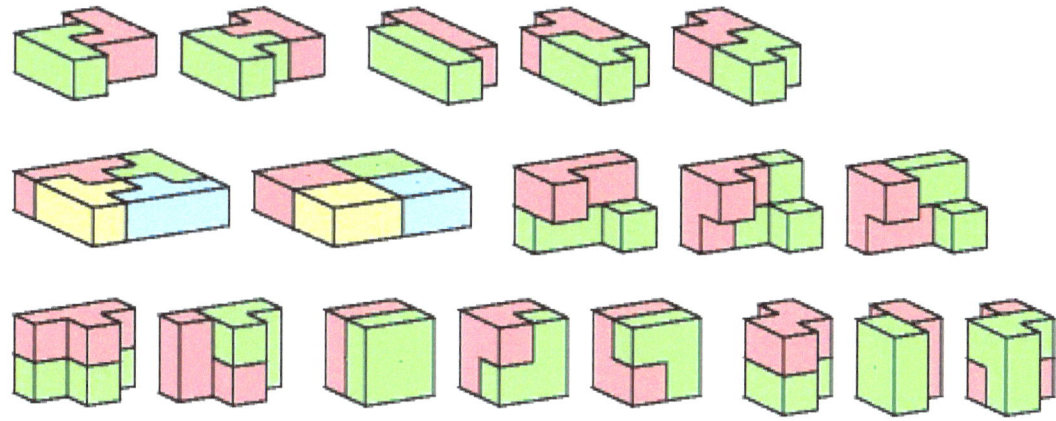

Tetracubes: 2D vs 3D

WHAT YOU WILL NEED

- Physical tetracubes (3D printed or purchased in advance)
- Clues for tetracube design
- 3D printer
- Method of saving student files

WHAT EACH STUDENT NEEDS

- Computer with SketchUp
- Tetracubes activity handout
- Sketching worksheet
- Pencil

OVERVIEW & PURPOSE

In this lesson, students will explore connections between 3D objects and 2D representations of those objects. Students will use 2D views of 3D tetracubes to construct the shape in three dimensions. They will also create a 3D design and represent the design with 2D sketches.

MATH COMMON CORE STANDARDS

CCSS.MATH.CONTENT.6.G.A.4

Represent three-dimensional figures using nets made up of rectangles and triangles, and use the nets to find the surface area of these figures. Apply these techniques.

CCSS.MATH.CONTENT.7.G.B.6

Solve real-world and mathematical problems involving area, volume and surface area of two- and three-dimensional objects.

CCSS.MATH.CONTENT.8.G.A.2

Understand that a two-dimensional figure is congruent to another if the second can be obtained from the first by a sequence of rotations, reflections, and translations; given two congruent figures, describe a sequence that exhibits the congruence between them.

A New Dimension of Mathematics with 3D Printing: **Grades 6 - 8**

Tetracubes: 2D vs 3D

THE ACTIVITY
Lesson Duration: 50 minutes

Thinking and Sharing

Tell students that today they will use SketchUp to view 3D objects from different angles. Note that designs do not appear the same from when observed from different views. For example, have the students consider an "upright" cylinder. Ask:

- What would this cylinder look like from the top? That is, what would it look like if we were in the sky looking down at it? (Answer: A circle)
- What would the cylinder look like from the side? (Answer: A rectangle)

Remind students of the Orbit tool in SketchUp, which allows the user to view designs from different angles. There are also standard Camera views in SketchUp, which are specific viewing options that can be selected. To introduce this tool to students, first have them quickly open a design of their choice from the SketchUp warehouse, such as a house or a superhero. This is a database of designs incorporated into the SketchUp software. Students should choose to load the design directly into their SketchUp file. They will try out the different viewing options on this object. To select a certain view, go to Camera on the top toolbar, and then Standard Views. From there students may select one of the following view options: Top, Bottom, Front, Back, Left, Right, Iso. The students may experiment for a couple minutes with these different views.

Tetracubes: 2D vs 3D

Students will first examine different views of tetracubes, providing a definition of a tetracube as well. Note that tetra means four, so a tetracube is a 3D object composed of four cubes that are connected. To further explain this concept, explain that tetracubes are three dimensional extensions of the 2D shapes composed of four connected squares seen in the game of Tetris.

To create tetracubes in SketchUp, students must first create and replicate cubes, which can be stacked together to form various tetracubes. If the students do not already know how to create and replicate cubes in SketchUp, explicitly go over the following steps as the students follow along on their own computers.

Step 1: Create a rectangle with a length and width of 1 unit. (Note that the SketchUp template is in millimeters, so to create a millimeter unit cube, enter "1, 1" in the measurement box. To create cubes with some other unit side length such as "1cm, 1cm." However, the measurement here is not very important, as long as the student uses the same measurement for each side of the cube.)

Step 2: Use the Push/Pull tool to extrude the square, and enter the same measurement as before (for example, 1cm) so that a cube is created.

Step 3: Make the cube a "group" so that the shape stays fixed. Then replicate the cube using copy and paste options. To do this, drag the cursor over the shape and let it go to select the entire cube. The entire cube should be highlighted in blue. Then use these keyboard commands:

A New Dimension of Mathematics with 3D Printing: **Grades 6 - 8**

Tetracubes: 2D vs 3D

- Group - Ctrl + G
- Copy - Ctrl + C
- Paste - Ctrl + V

Step 4: Replicate as many cubes as desired. These cubes can then be moved around to create designs.

Have students ask their partner, neighbors, a "class expert," or the teacher for help if they do not understand how to properly create or replicate a cube.

The students will now complete a tetracube activity that will challenge them to translate 2D pictures into a 3D design. Each student will be given a "clue." These clues are 2D representations of a tetracube, as seen from the top, front, and side.

On Their Own

The students must design the tetracube so that the 2D representations on their clue perfectly match their 3D design in SketchUp. An example is shown on the left.

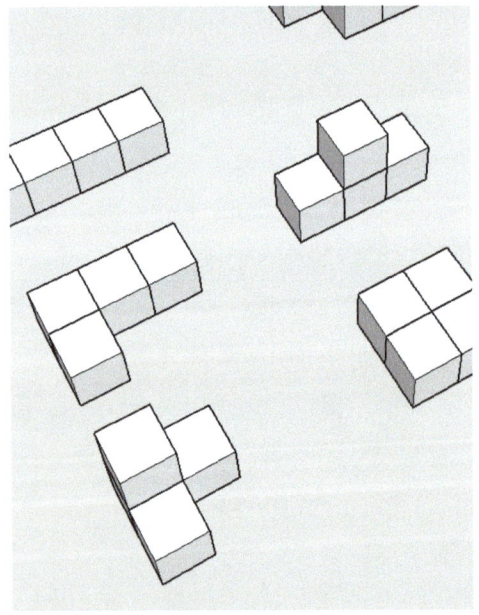

Tetracubes designed in SketchUp

Start by assigning one clue to each student. Once they have correctly designed that tetracube, have the student raise their hand to receive a different clue. Aim for most students to design about three tetracubes. Advanced students may get through more clues, while other students may get stuck on a clue. Alternatively, students could all be given the same clues in the same order. However, differing the level of difficulty for students is a strong form of differentiation. Assign students different clues based on their designing and spatial reasoning abilities. For example, students who are going through the clues quickly and successfully should be assigned more complicated tetracubes to design, such as "corkscrew"

A New Dimension of Mathematics with 3D Printing: **Grades 6 - 8**

Tetracubes: 2D vs 3D

tetracubes. Students who are struggling may be assigned less complicated tetracubes. It may be helpful to assign the same clue to groups or pairs of students so they may work together, especially at first.

As students design, some questions to ask are:

- How do you know your design is right?
- What tools could be useful to verify your design?
- How could you change your design to match the picture on your clue?

Remind students that they may not get the design perfectly right the first time, and that this task requires trial and error. They should continue to move the cubes around until they get a design that works.

Also, remind students to take advantage of the Orbit tool as well as the different camera viewing options that they practiced earlier. These are especially useful for students to use to verify that their designs match their clue once they think they are done.

Thinking and Sharing

- When the students are finished designing, have them again examine the physical tetracubes. Ask the students: Do any of these tetracubes look like the ones you designed?

Alternatively, the students could select one of their designs to be printed and could examine these designs at a later time.

Tetracubes: 2D vs 3D

Students may also share their designs and explain how they know it matches their clues.

Encourage discussion about the difficulties of going from 2D representations to actual 3D objects.

On Their Own

The students will now complete an activity that is the reverse of the previous activity; that is, they will create a 3D design and then sketch 2D representations of the design from different angles.

Give students 5 to 10 minutes to design anything of their choosing. The object should be made out of basic 3D objects they have learned about, such as prisms, cylinders, etc. The students should use no more than 10 of these shapes. Alternatively, the object should be made solely of unit cubes. These objects may be printed later for the students to keep, but remind students that the design must all be one piece in order to print and that the color of their printed object will not match the color of the design in SketchUp.

Next, have students sketch their object from different views (top, bottom, left, right, front, back) on a piece of paper. Remind students that they can use the Orbit tool and also the Camera standard views. Students will then trade papers with a partner. They will then try to design their partner's shape in SketchUp using the 2D sketches. After a certain amount of time, have students look at their partner's actual design to see if they were able to properly replicate it.

Tetracubes: **2D vs 3D**

Thinking and Sharing
Questions to discuss as a class or in groups:

- Did your design properly match your partner's design?
- Was it harder for you to go from a 3D design to 2D pictures or 2D pictures to a 3D design? Why?
- What relationships are there between 2D and 3D objects?
- Why is it important to be able to go from 2D to 3D objects?
- How does the computer software help us understand properties of shapes?
- How can the 3D printer help us see these relationships?

Remind the students that being able to translate between 2D representations and actual 3D objects is a familiar skill in the real world. For example, we do this when examining a map of a city or even taking pictures on our phone.

3D printers can help to emphasize the distinction between 2D and 3D objects. A 3D object will 3D print because it has volume; a 2D object will not print because it does not have volume, so there is nothing for the 3D printer to print.

Tetracubes: 2D vs 3D

FOR THEIR PORTFOLIO

- The student's tetracube or creative design(s)
- The student's sketches of their partner's construction from multiple views

GOING FURTHER

- Assign more complex tetracubes or a higher quantity of tetracubes to advanced students.
- On the second activity, encourage more advanced pairs of students to create designs that will challenge their partner once they trade papers.
- This lesson could potentially relate to a lesson in art class on points of perspective.

TEACHER TALK

- This lesson has students translate from 2D to 3D (the first activity, with tetracubes) and then from 3D to 2D (the second activity, with their own design), which helps to develop their spatial reasoning skills.
- This lesson is a good precursor to a lesson on cross sections, because often the top view of an object is the same as the cross-section of the shape when slicing it horizontally.

Tetracubes: 2D vs 3D

SKETCHING WORKSHEET

Name _____

Sketch your design from different views:

Front	Back
Left	Right
Top	Bottom

A New Dimension of Mathematics with 3D Printing: **Grades 6 - 8**

Tetracubes: 2D vs 3D

CLUE SET 1

Name _____

Design this tetracube:

Front:

Top:

Side:

A New Dimension of Mathematics with 3D Printing: **Grades 6-8**

Tetracubes: 2D vs 3D

CLUE SET 2

Name _____

Design this tetracube:

Front:

Top:

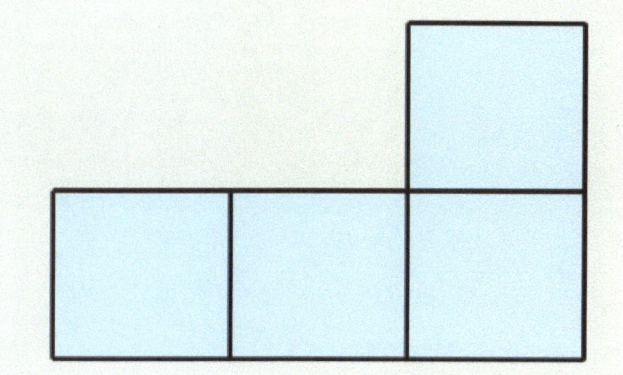

Side:

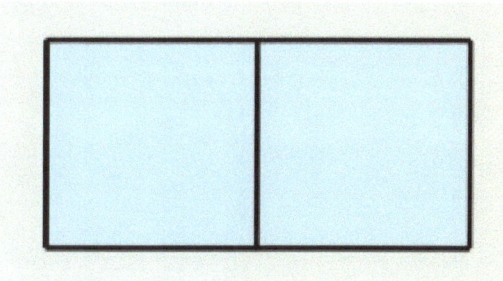

Tetracubes: 2D vs 3D

CLUE SET 3

Name _____

Design this tetracube:

Front:

Side:

Top:

A New Dimension of Mathematics with 3D Printing: **Grades 6 - 8**

Tetracubes: 2D vs 3D

CLUE SET 4

Name _____

Design this tetracube:

Front:

Top:

Side:

A New Dimension of Mathematics with 3D Printing: **Grades 6 - 8**

Tetracubes: 2D vs 3D

CLUE SET 5

Design this tetracube:

Front:

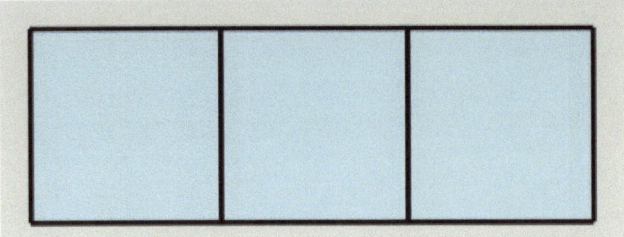

Top:

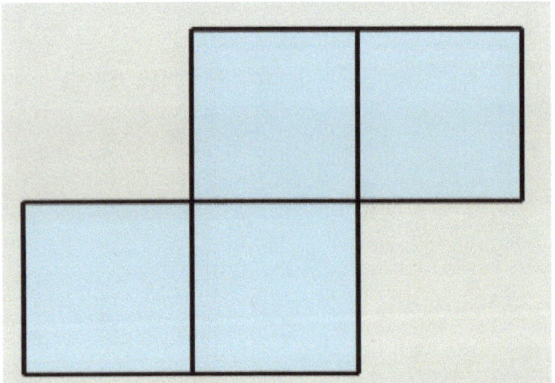

Side:

Tetracubes: 2D vs 3D

CLUE SET 6

Name _____

Design this tetracube:

Front:

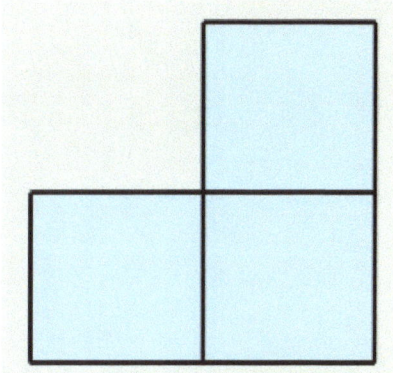

Top:

Right:

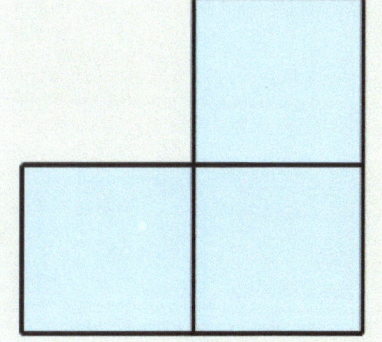

A New Dimension of Mathematics with 3D Printing: **Grades 6 - 8**

www.ingramcontent.com/pod-product-compliance
Lightning Source LLC
Chambersburg PA
CBHW051153220526
45473CB00003B/762